Brooklands Books

★ A BROOKLANDS ★
'ROAD TEST' LIMITED EDITION

BMW Z3 & Z3M

Compiled by
R.M.Clarke

ISBN 1 85520 4789

BROOKLANDS BOOKS LTD.
P.O. BOX 146, COBHAM,
SURREY, KT11 1LG. UK

ACKNOWLEDGEMENTS

For more than 35 years, Brooklands Books have been publishing compilations of road tests and other articles from the English speaking world's leading motoring magazines. We have already published more than 600 titles, and in these we have made available to motoring enthusiasts some 20,000 stories which would otherwise have become hard to find. For the most part, our books focus on a single model, and as such they have become an invaluable source of information. As Bill Boddy of *Motor Sport* was kind enough to write when reviewing one of our Gold Portfolio volumes, the Brooklands catalogue "must now constitute the most complete historical source of reference available, at least of the more recent makes and models."

Even so, we are constantly being asked to publish new titles on cars which have a narrower appeal than those we have already covered in our main series. The economics of book production make it impossible to cover these subjects in our main series, but Limited Edition volumes like this one give us a way to tackle these less popular but no less worthy subjects. This additional range of books is matched by a Limited Edition - Extra series, which contains volumes with further material to supplement existing titles in our Road Test and Gold Portfolio ranges.

Both the Limited Edition and Limited Edition - Extra series maintain the same high standards of presentation and reproduction set by our established ranges. However, each volume is printed in smaller quantities - which is perhaps the best reason we can think of why you should buy this book now. We would also like to remind readers that we are always open to suggestions for new titles; perhaps your club or interest group would like us to consider a book on your particular subject?

Finally, we are more than pleased to acknowledge that Brooklands Books rely on the help and co-operation of those who publish the magazines where the articles in our books originally appeared. For this present volume, we gratefully acknowledge the continued support of the publishers of *Autocar, Automobile Magazine, Autosport, Car and Driver, Car Magazine, Motor Sport, Motor Trend, Road & Track, Road & Track Specials, Sports Car International, Top Gear, What Car?* and *Wheels* for allowing us to include their valuable and informative copyright stories.

R.M. Clarke.

CONTENTS

4	BMW Z3	*Car and Driver*	Sept		1995
6	Licence to Kill	*Autocar*	June	14	1995
10	BMW Z3 Sub-£20,000 Dreamwagen	*What Car?*	Jan		1996
14	BMW Z3 Road Test	*Road & Track*	Jan		1996
21	Film Star's Variety Performance	*Autosport*	Mar	7	1996
22	Rule of Three Road Test	*Sports Car International*	Feb		1996
26	Porn Jerry - BMW Z3 vs. MGF Comparison Test	*Wheels*	June		1996
30	BMW Z3 - 2.8 Road Test	*Road & Track Specials*			1997
36	Showdown! - BMW Z3 2.8 vs. Porsche Boxster vs. Mercedes-Benz SLK Comparison Test	*Automobile Magazine*	Mar		1997
43	Straight Six Pushes BMW Z3 into the Lead	*Motor Sport*	Feb		1997
44	Dial M for Murder - M Roadster	*Autocar*	Mar	5	1997
48	BMW Z3 2.8 Road Test	*Motor Trend*	Mar		1997
51	BMW Z3	*Motor Trend*	Aug		1996
52	Long-Term BMW Z3	*Car and Driver*	Aug		1997
56	Nose to Tail - BMW Z3 2.8 vs. Porsche Boxster Comparison Test	*Autocar*	Sept	3	1997
64	BMW Z3 - M Roadster Road Test	*Autocar*	Jan	28	1998
68	M-Power for Thrills	*Car Magazine*	Mar		1998
69	BMW M Roadster	*Autocar*	Dec	17	1997
70	Road Hunks - BMW M Roadster vs. Corvette Convertible vs. Porsche Boxster Comparison Test	*Car and Driver*	Mar		1998
78	BMW M Roadster	*Automobile Magazine*	Mar		1998
81	BMW Z3	*Road & Track Specials*			1998
82	Mach is MO - BMW M Roadster vs. TVR Chimaera 5.0 Comparison Test	*Top Gear*	Mar		1998
87	BMW Z3 1.9i and BMW Z3 2.8	*Autocar*	Dec	17	1997
88	Scary Spice vs. Porsche Spice - BMW Z3 M Coupé vs. Porsche 911 Carrera Comparison Test	*Autocar*	July	1	1998
92	BMW M Roadster Road Test	*Road & Track*	Mar		1998

PREVIEW

BMW Z3

A first look at this new American-made Bimmer sports car, which may arrive in time for Christmas.

BY PETER ROBINSON

Eyeball these photographs. They're the only chance you have to see BMW's long-awaited Z3 roadster until December, when the movie *Golden Eye* opens.

To the dismay of the British Empire, the allure of this BMW shot down Aston Martin's hopes of casting the new DB7 as James Bond's exotic mode of transport in his latest saga. Instead, he'll drive an American-made, German-designed ragtop. Well, everything's gone to hell since Connery bailed out anyway.

BMW will also upstage by several months the debuts of two other much-previewed German sportsters. The Z3 will have its premiere at next January's Detroit auto show; a few weeks later, this South Carolina–built rival for Mazda's Miata will go on sale in both the U.S. and Europe. Porsche now plans to unveil the production Boxster at the Geneva show in March 1996 but won't start taking your money until September. Mercedes plans a June launch for its SLK, at least in Europe.

Meanwhile, these are the first official photographs of the Z3, BMW's "affordable" roadster. Affordable? BMW's target price for the U.S.—easily the most important market for the new car—is $25,000. Unless there's a dramatic shift in exchange rates, that means the base 113-hp Z3 will cost around $7000 more than the now six-year-old MX-5 Miata.

So sensitive is the price of the Z3 that, unofficially, Rover management asked BMW—remember, BMW now owns Rover—to raise the price of the Z3 in Britain. Rover's problem is that the new 143-hp MGF, the reborn version of the classic MG (which won't be sold in the U.S.), is expected to cost about $29,500, and the 138-hp BMW Z3 (which is the only version to be sold in the U.K.) should run just $800 or so more.

But with BMW's plant in Spartanburg, South Carolina, set to build 30,000 Z3s a year, Britain is too important a market to restrain artificially. Munich's decision: The MGF and the Z3 will compete head-on.

Unlike the mid-engined MGF, the Z3 follows a more traditional plan. It's built on the 3-series hatchback platform, with the old-style semi-trailing-arm rear suspension, and its rear wheels will be driven by either of BMW's two front-mounted 1.8-liter four-cylinder engines. For most markets, the entry-level engine is the eight-valve four taken directly from the European 318i; it produces 113 horsepower at 5500 rpm and 124 pound-feet of torque at 3900 rpm. Those who prefer the more

4

BMW Z3

sporting character and spirited acceleration—0 to 60 mph in close to 8.0 seconds—of the 16-valve DOHC version will have 138 hp at 6000 rpm and 129 pound-feet of torque at 4500 rpm to play with.

The long-nose/short-tail profile so obvious in the photographs isn't just a styling gimmick. BMW's brilliant 193-hp, 2.8-liter, all-aluminum in-line six also fits under the hood. Expect it late in 1997, to give BMW a model capable of matching the supercharged Mercedes-Benz SLK and the six-cylinder Porsche Boxster.

Buyers of the Z3 will choose between a five-speed manual gearbox and an optional GM four-speed automatic. Later, ZF's five-speed auto will also be added to a long list of options that will eventually include a hard top, a power-operated roof, and air conditioning.

"Go-kart like" is the message getting out about the Z3's handling. To improve agility and provide more direct steering, the ratio of the 318ti's power steering was reduced by about 20 percent for the Z3. Spring and damper rates are firmer and the diameter of the anti-roll bars larger, while the track is increased slightly by 0.28 inch at the front and 0.55 inch at the rear. A 4.3-inch reduction in length (compared with the 318ti) to 161.4 inches helps, but the most significant dimensional change is a nearly 10-inch reduction in the 3-series' 106.3-inch wheelbase to give the Z3 a wide-track stance.

In contrast to the kart-ish Z3 four-cylinder, the Z3 2.8 will feature a more boulevard-quality ride, with softer springs, dampers, and anti-roll bars. A sport suspension will be available, but it too will be softer than the base Z3's.

The Z3 is bigger than the Miata. The designers have exploited the added space to ensure that the BMW is both more versatile and more practical. The cockpit offers plenty of bins for minor luggage, and the trunk is sensibly shaped and relatively large for this kind of car. Nonetheless, the Z3 is a pure two-seater. If you want to carry a third person, you'll have to wait for the 1997 Z3 Hatchback, which will have a single, transversely mounted seat.

Weight? No official figure yet, but the same insiders who say the Z3 will have a rigid and highly crashworthy body structure guess that it will weigh about 2600 pounds, virtually the same as the 318ti Compact.

BMW's designers have opted for a look that combines just the right degree of retro-styling—those side vents are supposed to remind us of the 1956–59 BMW 507—with plenty of contemporary detailing. Prominent wheel arch bulges are filled by 16-inch alloy wheels.

Customers can choose their own level of exclusivity. Beyond a wide range of exterior and interior colors and upholstery materials, BMW will offer a choice of optional aerodynamic components or exterior chrome trim, so the buyer can go with either a modern or a traditional look.

BMW has learned the lessons of the high-priced ($47,000), limited-volume Z1 sports car that struggled to sell just 8000 examples between 1989 and 1992. The Z3 represents the opposite approach. In fact, BMW sees its new sports car taking over from Mazda's MX-5 Miata as the world's best-selling ragtop roadster. •

Vehicle type: front-engine, rear-wheel-drive, 2-passenger, 2-door convertible
Estimated base price: $25,000–28,000
Engines: SOHC 8-valve 1.8-liter 4-in-line, 113 bhp, 124 lb-ft; DOHC 16-valve 1.8-liter 4-in-line, 138 bhp, 129 lb-ft
Transmission .... 5-speed, 4-speed automatic with lockup torque converter
Wheelbase .... 97.5 in
Length .... 161.4 in
Width .... 66.9 in
Curb weight .... 2600 lb
C/D projected performance:
Zero to 60 mph .... 8.0–9.0 sec
Standing ¼-mile .... 16.0–18.5 sec

BMW Z3 ROADSTER

A year from now, when BMW's new Z3 roadster goes on sale in the UK, the MGF will confront its toughest rival. Ironically, the competition comes from its own ranks and, wearing the white and blue propellor badge, MG knows it could hardly be stronger.

The Z3, officially unveiled here for the first time, is BMW's long-awaited, affordable roadster. Affordable? How does £19,000 sound? Unless there's a dramatic shift in exchange rates, the 140bhp 1.8-litre Z3 will cost only £500-1000 more than the 143bhp VVC MGF.

Unofficially, Rover bosses asked BMW to raise the Z3's UK price. But with BMW's plant in Spartanburg, South Carolina, set to build 30,000 Z3s a year, the UK is too important a market to rig. Munich's decision: the MGF and Z3 will compete head on.

So MG must simply hope currency fluctuations will force BMW to increase the price. For the moment the target is still £19,000. The problem is Rover's deal with Mayflower, which engineered and now produces the mid-engined MGF's body, doesn't allow any flexibility on pricing.

Unlike the mid-engined MGF, BMW has gone the traditional route for the Z3 roadster. Built on the 3-series Compact platform, with the old-style semi-trailing arm rear suspension, the Z3's rear wheels will be driven by either of two front-mounted, 1.8-litre fours.

For the UK, the eight-valve 115bhp four from the 318i saloon has been ignored. BMW believes UK buyers will prefer the more sporting character and spirited acceleration – 0-60mph in close to 9.0sec – of the 16-valve twin-cam version. Of course, it also makes selling the 118bhp, £16,000 MGF that much easier. Coincidence? Make up your own mind.

BMW's brilliant 193bhp 2.8-litre in-line six also fits and will be added to the range, probably in 1997, to give BMW a model capable of taking on the forthcoming supercharged Mercedes SLK and six-cylinder Porsche 986 Boxster. Buyers will choose between a five-speed manual gearbox or optional GM four-speed auto. Later, ZF's five-speed auto will also be added to a long list of options that will eventually include a hard-top, power roof and air conditioning.

Those who've driven the Z3

Licence TO KILL

At £19,000, BMW's Z3 could make the MGF an endangered species, says Peter Robinson

BMW Z3 ROADSTER

In UK next year with 1.8 16-valve power and £19,000 price. Hot 2.8-litre version planned for '97

Z3 based on Compact platform. Styling mixes modern detailing with design cues of past (right)

claim it has kart-like handling. The ratio of the power steering has been reduced by about 20 per cent to give more directness. Spring and damper rates are firmer and the diameter of the anti-roll bars larger, while the tracks are increased slightly by 7mm at the front, 14mm at the rear. A 110mm reduction in length, to 4100mm, improves the Z3's stance, though the long 2700mm wheelbase stays. That makes the Z3 bigger than both the MGF and Mazda's MX-5.

The added space means the BMW is both more versatile and practical. The cockpit offers plenty of bins for minor luggage, while the boot is sensibly shaped and said to be noticeably bigger than those of all existing rivals. Yet it remains a two-seater only. If you want to carry an extra passenger you'll have to wait for the 1997 Z3 hatchback that adds an extra, transversely mounted seat.

Weight? No official figure yet, but insiders say that giving the Z3 a rigid and crashworthy body structure has pushed it up to about 1175kg, virtually the same as the 318i Compact's.

BMW's designers have opted for a look that combines just the right degree of retro styling – those side vents are supposed to remind us of BMW's 507 roadster of the '50s – with plenty of contemporary detailing. Prominent wheel arch bulges are filled by 16in alloys.

Beyond a wider-than-normal choice of exterior and interior

> **'BMW sees the Z3 taking over from Mazda's MX-5 as the world's best-selling roadster'**

Licence TO KILL

colours and upholstery materials, you'll be able to order aerodynamic components and add chrome to the exterior. In effect, you'll be able to decide between body options that provide either a traditional or modern appearance.

Before the Z3 goes on sale in the UK, you'll see it in action in December. The Z3 is James Bond's wheels in the movie *Golden Eye*. Its first public appearance will be at January's Detroit motor show, just before production of left-hand-drive cars begins.

BMW has learnt the lessons of the high-priced, limited-volume Z1 roadster from the '80s, which sold just 8000. The Z3 takes the opposite approach; in fact, BMW sees it taking over from the MX-5 as the world's best-selling roadster. Even at the expense of Rover, whose MGF won't be sold in the US. ●

Newcars

BMW Z3
On sale July
Expected price Below £20,000
Category Roadster

BMW Z3
Sub-£20,000 dreamwagen

Desirable roadster combines bold looks with rear-wheel drive and a tantalising chassis

The excitement generated by the first all-new MG roadster in 30 years has barely had time to cool. Yet already there's a challenge to its authority – from none other than MG's owner, BMW.

The Z3, a two-seater roadster built in the US, will arrive in the UK in July as a one-model-only 1.9-litre 16-valver, packing 140bhp. With an expected price of less than £20,000, it will sit close enough to the £18,500 145bhp MGF 1.8i VVC to compete directly.

BMW evidently thinks the Z3 and MGF will happily co-exist. They are sufficiently different in character and looks, if not in their philosophy of what a new-age roadster should be.

The Z3 is more traditional than the mid-engined MG. It has the front-engine/rear-drive layout of all BMWs, and this influences its shape – long-nosed and short-tailed with a steeply angled screen and far-back cockpit.

It looks a wide car, emphasised by the muscularity of its front wheelarches and broad nose, containing the familiar BMW kidney grille that grows out of a couple of creases in the bonnet. At the rear, a wedge has been carved from the boot lid to create a space for a high-level brake light.

No new roadster is complete without those design details meant to stimulate nostalgic yearnings. The most obvious on the Z3 are two lovely metal-and-mesh air outlets, complete with BMW roundels, gracing the panels behind the front wheels.

'Above all, you'll thoroughly enjoy driving and owning it'

Sculpted nose is reminiscent of BMW sportsters from the '50s

They may evoke memories of the BMW 507 of the late '50s, but serve no practical function.

The cabin looks minimalist, though you can dress it up with wood and leather. We'll soon know which trims and electrical options BMW plans to make available in the UK.

The seats hug you just firmly enough and the belts are routed to fit snugly over your shoulder. The instruments will be familiar to anyone who has ever driven a BMW, but the dash is straight, not curved towards the driver as in sister models.

The passenger area is clever, with an overhanging facia to boost legroom and provide space for tall occupants to sit upright in comfort, but the steering wheel is a little too large and high-set. Few people will complain that the cabin is cramped and there is ample stowage, including a couple of lockable console boxes and a net to house maps that lives in the passenger footwell. It also has a good-sized boot.

The single-layer hood, clamped to the screen by two latches, arcs backwards and contains a decent-size zip-out rear window. It slots neatly into a well behind the seats and is covered by a stud-fastening leatherette tonneau, though in time you'll be able to buy a hard top and panel with a couple of speedster humps.

FIRST DRIVES
BMW Z3

A mesh wind-deflector, like that offered in the 3-series cabriolet, is another option, though the car is reasonably free from buffeting when driven with the side windows lowered.

The car's structure feels tremendously stiff thanks to the strength of its screen frame, doorposts and sills. Bracing in the engine bay (and between floorpan, bulkheads and roof compartment), plays its part, too. It feels as safe as a roadster can ever do, with side-impact beams, seatbelt pretensioners and grabbers, twin airbags and body-retaining seats all offering reassurance.

So much stiffening and safety strengthening does it few favours in the weighing enclosure: the Z3 is heftier than either the MG or the Fiat Barchetta and – on paper, at least – slower.

The engine, an enlargement of the 1.8 16-valve unit from the 3-series coupé and Compact, gains no more power, though reductions are claimed in both fuel consumption and emissions, and there's a useful increase in pulling power at low-to-medium revs to make it more driveable.

It still doesn't come alive until it's spinning at 3000rpm or more, but you don't have to push it hard unless you're crossing the Alps or trying to pass a pantechnicon. The short, stabbing action of the ▷

Newcars

You can add wood and leather by choosing from option packs. Cabin is roomy, with good storage space

Any BMW driver would instantly feel at home here. Unlike 3-series saloon, dash is flat and wheel is large for a roadster

◁ gearlever makes shifting a joy, though you can order a four-speed auto gearbox as an option.

Will there be six-cylinder Z3s, using BMW's new 2.3- or 2.8-litre units? These engines fit and the Germans say they will install them in the Z3 if there's enough demand. We bet there will be.

Really, though, roadster motoring is more about the joy of travelling than the speed, and the 1.9-litre engine should prove sufficiently brisk and relaxed for many owners.

BMW has not only made the 1.8 engine bigger, but has also changed its internals and cylinder head substantially, making it smoother and sweeter-sounding. The gritty, clattery noises of the old 1.8-litre have disappeared and it is quiet in normal running, though there's a pleasing bass rasp when you rev it.

In character, the Z3 sits somewhere between a Mazda MX-5 and an MGF – less raw and overtly sporty than the Japanese car; sharper but every bit as refined as the British one. Put your hand on the Z3's gearlever as you drive along and there's no hint of vibration through the driveline, while the level of bump suppression from the suspension, like that of the Hydragas-sprung MGF, shows that it is possible to drive an open car without suffering a kart-like ride.

Broken tarmac can induce minor shivers and there's something harder (and louder) when crossing potholes or raised ridges in the road, but its ride still outstrips that of many closed cars.

Most of the Z3's underpinnings are taken from the BMW Compact, though its track is wider, its suspension has been made firmer, it is given quicker steering and bigger brakes, and perches on a shorter wheelbase.

It also has a low centre of gravity, weight distribution only fractionally biased towards the front, and sits on 15in, 16in or even 17in wheels wearing substantial tyres, so it's almost impossible to unstick.

It can't thrill like an MX-5, but it is easier to live with. Stretched to the limit, the front tyres are the first to move gently off-line, but the rears stay firmly anchored, even under hard acceleration out

Z3 is similarly sized to its main rivals, MGF and MX-5. Expect to find it on enthusiasts' shopping lists beside MGF 1.8 VVC

FIRST DRIVES
BMW Z3

of a tightish, damp bend. It takes violent provocation to unsettle the back end, and there's even the option of stability/traction control to counter this electronically.

The standard power steering is geared to need only three turns between the lock stops: this is one of the Z3's delights, giving a more direct link between driver and front wheels than you will find in other 3-series cars. There's still a half-inch of indecision in the wheel's movement when pointing straight ahead, and that makes the car a mite unsure on a fast, bumpy road, but when you

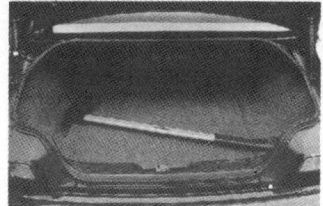

Practical boot is a useful bonus

attack corners its reaction is accurate, swift and certain. Anything sharper might have been too nervous.

There's nothing here that the owner of a coupé or hot-hatch – the people likely to form the core of roadster buyers – would find objectionable. It's about as safe as it's possible to make an open car, and as secure. It has BMW's acclaimed immobiliser system, and even the space-saving spare wheel, cradled beneath the car, unlatches through the boot floor to prevent theft.

You can live with the Z3, personalise it with any number of option packs and, above all, thoroughly enjoy driving and owning it. Exactly the same can be said of the MGF, of course. We won't know which is the better until we pitch one against the other in a few months' time, but the outcome will be very, very close. Choose either, and you simply can't go wrong. ∎

BMW Z3	
Engines	1895cc 4cyl 16v, 140bhp
Trim levels	One, with option packs
Touring mpg	34.4
Performance	127mph, 0-62 9.5sec

FIRST GLANCE

Epitome of the '90s roadster: great to look at, easy to live with, pleasing rather than thrilling to drive

VERDICT ●●●●

13

1996 BMW

Z3

Bred in Bavaria, built in South Carolina and bound for stardom

BY KIM REYNOLDS
PHOTOS BY GUY SPANGENBERG

ROAD & TRACK ROAD TEST

ALTHOUGH BMW CRINGES at even whispered comparisons between its Z3 and the Mazda Miata, let me start explaining the Z3 with the following contrast: In 1990, with Mazda's public relations department hardly having to lift a finger, the new Miata created a media sensation. Major newspapers ran front-page stories on "Miata mania." Evening news broadcasts sent cameras with satellite links out to Mazda dealers swamped with check-waving hopefuls.

Six years later, it is the Z3 that is elbowing its way into our mass consciousness, but notice a qualitative difference. Now it's via avenues like swanky Nieman-Marcus that included a decked-out Z3 in their recent catalog or the latest James Bond movie, *GoldenEye*, in which good old Q has slid 007 behind a Z3's wheel. Unlike the Miata phenomenon, the Z3's media presence isn't spontaneous combustion; it's engineered awareness. Carefully planned. BMW is an engineering, plan-oriented company that isn't about to just throw the Z3 into its dealerships and hope for a big splash. Maybe they brood over details longer in BMW's Munich than in Mazda's Hiroshima; either way, you end up feeling that fine-tooth-comb difference driving the Z3 on the road. Think of a Miata built by...well, BMW (with an extra $10,000 to spend), and you'll begin to understand the Z3.

15

Dimensionally, the Miata and the Z3 are certainly comparable cars. The Z3 is longer (by 3.1 in.) and slightly wider (by 0.7 in.), but what stands out most is that while their tracks are virtually identical, the Z3's wheelbase is a substantial 7.1 in. longer. Another notable difference is weight with the Z3 pushing the scale dial up an extra 400 lb. (based on BMW's figures) and becoming roughly a 2700-lb. automobile. Partly this is because of obviously heavier-grade materials and the sort of extra features that befit its price. But clearly a lot of that weight resides in its highly effective structure.

On the road, these three factors—a relatively long wheelbase, additional weight and greater stiffness—add up to a ride quality that's positively plush for a car that's still really small. There's no choppiness to speak of, zero cowl shake from what I could tell; and even big potholes are largely absorbed. It's the best-riding, most structurally robust, small open sports car in my memory.

The source of that ride, its suspension as well as its engine and transmission are familiar friends if you've had a chance to read about or drive a 318ti; they're all the same components except for an extra 99 cc in displacement. Up front is the 3-Series' astonishingly well-developed MacPherson struts and lower lateral arms, while at the rear are the ti's revived semi-trailing arms—a point BMW seems to be slightly defensive about. During the Z3's introduction, the spokesman described these trailing arms, not as the 318ti's, but as being identical to "the first-generation M3's" and then briefly stared at us, the attending (perhaps offending) press. As I'll get to in a moment, there's nothing performance-wise about these semi-trailing arms to be defensive about. From a packaging standpoint, they allow for a fairly reasonable trunk cavity that would have been impossible had the multilink system (employed by the rest of the 3-Series range) been used.

Underhood is the likable BMW 1.9-liter 4-valve 4-cylinder that puts out 140 bhp at 6000 rpm and 133 lb.-ft. of torque at a usefully low 4300 rpm. Power to weight-wise, the Miata (its shadow never really leaves, does

It's a handsome car, Germanically muscular, and with some deeply sculptured details that I enthusiastically applaud.

it?) has a mathematical edge at 17.2 lb./bhp versus the Z3's 19.2. But in reality, the Z3 is the slightly quicker car from a stop because of its nearly equal torque-to-weight ratio (with its peak torque reached at 1200 less rpm) and its ability to spin its rear wheels up to a much higher speed.

Esthetically, the 1.9-liter has a smooth, businesslike voice all the way to its 6350-rpm redline. But unlike the 318ti—where engine note doesn't much matter—here, in a sports car, its lack of throb, burble and pop is more apparent. The best it does is rasp a little. What you want to hear from a roadster's tailpipe is the scat singing of Ella Fitzgerald, not the mellow sound of Mel Torme.

Another carping criticism I'd levy is with the Z3's steering—not its precision, weight and linearity, which are all BMW-excellent—but with its weak registering of road-surface irregularities. It's almost too good; I'd prefer the Z3's steering wheel to vibrate and flinch in my hands a bit more.

Drive the Z3 casually, and you'll be comfortable for probably as far as the trunk space will allow you to go. At six-foot-one, I had the seat in its fully aft position with the seatback pressed firmly against the rear bulkhead, but I felt perfectly comfortable. With the top down (it can be put back up from the driver's seat if your right arm is strong enough), wind buffeting is quite reasonable, and oddly I found it was minimized when the driver's window was down and the passenger's was up (better yet is an optional flip-up wind-blocking screen located behind the seats).

The first time I flicked the Z3 into a turn, its quick roll rate surprised me. Obviously the springs are soft and BMW hasn't fit very large anti-roll bars, but once you're accustomed to its roll angles, the Z3's handling balance and sheer flingability are positively acrobatic. Deliberately drive too hard into a turn, and it cautiously understeers. Lift and pump the throttle pedal, and the tail slides out with mind-reading obedience. Lap after lap around the skidpad, I could hold the Z3 at what felt like a 30-degree tail-out angle with the possibility of an unintentional spin never entering my mind.

But it was through the slalom cones that another dimensional aspect of the Z3 became apparent, and that's how

■ Those gills on each side of the Z3 give this new roadster a shark-like presence, and they also allow engine heat to escape.

■ A removable rollbar, available at a later date, will fit into the covered slots behind the seats (between them is a lockable storage bin). As is typical of BMW, the gauges are large and easy to read, and the engine bay has enough room for one of the company's noted inline-6s, which will soon find its way into the Z3. The cockpit, covered here in optional leather, is a bit larger than that of a Miata.

far astern you're sitting in the car. With your butt about one foot ahead of the rear tires (somewhat like a Lotus Seven), every slip of the rear end seems magnified, and at first I was anxiously correcting my way out of every little drift. But after a while, your brain seems to rewire itself to the new geometry and the amplified motions are recorded as fun, not fear.

When you see a Z3 on the road, I think you'll agree it's even better-looking than these pictures suggest. It's a handsome car, Germanically muscular, and with some deeply sculptured details that I enthusiastically applaud. The grille kidneys are spot-on, the headlight lenses clipped on both their tops and bottoms to give a menacing stare, and the 507-evocative side gills have a playfully reptilian look. Walk rearward from the Z3's nose and its scale almost seems to shrink, reducing to something like 7/8ths by the time you reach the rear bumper. The effect unequivocally declares it as a front-engine roadster. Although the Z3 will be offered in dark blue, green, red and aqua, I'd think its rippling shape is best seen in either of the two metallic colors, the silver or light blue (the James Bond color).

And now to the matter of price. A base Miata today costs $18,450 which is—click, click, click on the calculator—64 percent of the Z's $28,750 starting fee. Frankly, that seems like a lot for a sub-2-liter small sports car, but on the other hand, a BMW roadster

1996 BMW Z3

MANUFACTURER
BMW of North America, Inc.
P.O. Box 1227
Westwood, N.J. 07675

PRICE
List price................................ $28,750
Price as tested $32,599

Price as tested includes std equip. (dual airbags, ABS, air cond, cruise control, AM/FM stereo/cassette, central locking, foglights; pwr windows, mirrors & seat adjustment), leather interior ($1150), traction control ($1100), heated seats ($500), metallic paint ($475), dest charge ($570), luxury tax ($54).

0–60 mph	8.1 sec
0–¼ mi	16.1 sec
Top speed	116 mph*
Skidpad	0.91g
Slalom	62.2 mph
Brake rating	excellent

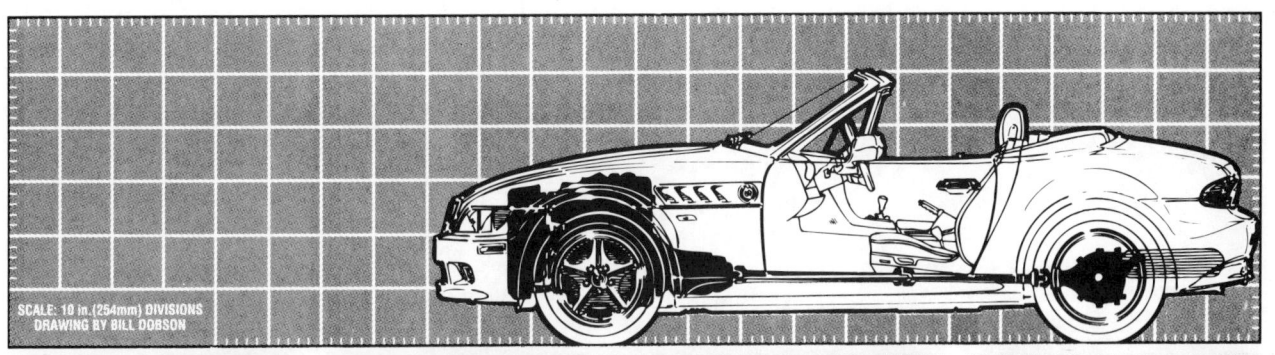

TEST CONDITIONS
Temperature .... 65° F
Wind .... calm
Humidity 32%
Elevation .... na

SCALE: 10 in.(254mm) DIVISIONS
DRAWING BY BILL DOBSON

ENGINE
Type........... aluminum block and head, **inline-4**
Valvetrain........... dohc 4-valve/cyl
Displacement.... 116 cu in./1895 cc
Bore x stroke....... 3.35 x 3.29 in./ 85.0 x 83.5 mm
Compression ratio........... 10.0:1
Horsepower
  (SAE)....... **140 bhp @ 6000 rpm**
Bhp/liter................... 73.9
Torque....... **133 lb-ft @ 4300 rpm**
Maximum engine speed.... 6350 rpm
Fuel injection... elect. sequential port
Fuel.... prem unleaded, 91 pump oct

CHASSIS & BODY
Layout...... **front engine/rear drive**
Body/frame............... unit steel
Brakes
  Front............. **11.3-in. discs**
  Rear.............. **11.0-in. discs**
  Assist type.......... vacuum; ABS
  Total swept area........ 407 sq in.
  Swept area/ton .... 291 sq in.
Wheels........... light alloy, **16 x 7J**
Tires........... Michelin Pilot MXM, **225/50ZR-16**
Steering............. **rack & pinion**, power assist
  Overall ratio............... 13.9:1
  Turns, lock to lock............ 2.9
  Turning circle............. 39.4 ft
Suspension
  Front......... **MacPherson struts**, lower L-arms, coil springs, tube shocks, anti-roll bar
  Rear......... **semi-trailing arms**, coil springs, tube shocks, anti-roll bar

DRIVETRAIN
Transmission.. **5-sp manual**

Gear	Ratio	Overall ratio	(Rpm) mph
1st	4.23:1	14.60:1	(6350) 31
2nd	2.53:1	8.73:1	(6350) 52
3rd	1.66:1	5.73:1	(6350) 78
4th	1.22:1	4.21:1	(6350) 106
5th	1.00:1	3.45:1	est (5690) 116*

Final drive ratio 3.45:1
Engine rpm @ 60 mph in 5th 2950
*Electronically limited.

GENERAL DATA
Curb weight............ **est 2690 lb**
Test weight .... est 2850 lb
Weight dist (with driver), f/r, %....... est 52/48
Wheelbase................... 96.3 in.
Track, f/r........ 55.6 in./56.2 in.
Length................... 158.5 in.
Width..................... 66.6 in.
Height.................... 50.7 in.
Ground clearance .... na
Trunk space .... 6.2 cu ft

MAINTENANCE
Oil/filter change ... 7500 mi/7500 mi
Tuneup .... 30,000 mi
Basic warranty..... 48 mo/50,000 mi

ACCOMMODATIONS
Seating capacity 2
Head room .... 37.0 in.
Seat width............ 2 x 19.0 in.
Leg room................. 43.0 in.
Seatback adjustment .... 35 deg
Seat travel .... 8.0 in.

INTERIOR NOISE
Idle in neutral .... 48 dBA
Maximum in 1st gear........ 77 dBA
Constant 50 mph .... 65 dBA
70 mph .... 73 dBA

INSTRUMENTATION
250-kph (155-mph) speedometer, 7000-rpm tach, coolant temp, fuel level

ACCELERATION
Time to speed	Seconds
0–30 mph	2.6
0–40 mph	4.0
0–50 mph	5.7
0–60 mph	8.1
0–70 mph	10.9
0–80 mph	14.5

Time to distance
0–100 ft....................... 3.2
0–500 ft....................... 8.7
0–1320 ft (¼ mi): 16.1 @ 84.5 mph

FUEL ECONOMY
Normal driving .... est 25.0 mpg
EPA city/highway ..... est 23/32 mpg
Cruise range........... est 312 miles
Fuel capacity............. 13.5 gal.

BRAKING
Minimum stopping distance
  From 60 mph.............. 118 ft
  From 80 mph.............. 211 ft
Control.................. excellent
Pedal effort for 0.5g stop......... na
Fade, effort after six 0.5g stops from 60 mph .... na
Brake feel .... excellent
Overall brake rating........ excellent

HANDLING
Lateral accel (200-ft skidpad)... 0.91g
  Balance.......... mild understeer
Speed thru 700-ft slalom ... 62.2 mph
  Balance....... moderate understeer
Lateral seat support........ excellent

Test Notes...

■ Around the skidpad the Z3 not only breaks 0.9g, but is finely balanced at almost any attitude. With stabs of throttle or brake, the Z3 can be placed—with confidence—exactly where you want it.

■ With stopping distances of 118 and 211 ft from 60 and 80 mph, the Z3 halts as well as it corners. BMW doesn't offer vented front discs, but we never noticed any sign of fade.

■ While the Z3's shifter isn't quite Miata-sublime, it's still very good—precise and with short throws. The pedals are ideally placed for rapid-fire heel-and-toe downshifts.

Subjective ratings consist of excellent, very good, good, average, poor; na means information is not available.

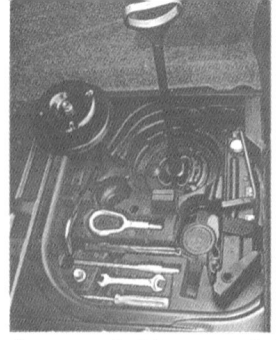

■ The Z3's spare tire resides below the trunk, and can be lowered via the special tool, at far left. For model year 1997, BMW will offer the Z3 with a removable hardtop.

for around thirty grand doesn't sound that bad either. Basically, it's one of those situations where if you have the money, you'd be silly not to go for it; and if you can't afford it, you'd be dumb to try. In other words, it's a sports car.

Which brings us to a final and almost overriding question: Does the Z3 herald the return of the traditional sports car (along with the upcoming Porsche Boxster and Mercedes-Benz SLK) or is it the last stand against a plethora of alternative sporty methods for getting around these days?

Clearly, sales of fast, pricey technology-rich sports/GT cars are in a frightful tailspin. But my guess is that the sports-car market isn't so much mortuary-bound as it is in the throes of a reincarnation of the classic, simple roadster theme, precisely the back-to-the-future scenario BMW foresees. The Z3's future looks really good.

Before the Z3

Because BMW has rarely built roadsters, most people probably don't connect the two. But the very few 2-seat open cars the Bavarian Motor Works has offered have been knockouts.

With its drape-like bodywork, twin leather straps over its hood and artistically minimal bodywork detailing, the 328 of 1936 was a styling classic. It was also a very good racing car, in part because of its engine's innovative solution to combining pushrod valve actuation with hemispherical combustion chambers, resulting in 80 bhp from a displacement of less than 2 liters.

During World War II, BMW resumed building aero-engines (as it had in WWI) and was subsequently destroyed by Allied bombing.

But by 1956, it was ready to roll out another roadster, the V-8 sedan-based 507. Styled by Count Albrecht Goertz (who later shaped the original Datsun 240Z), the beautiful 150-bhp 507 is often mistaken for a coupe because its removable hardtop is so well integrated.

More recently, the Z1, which bowed in 1986 as a showcase for BMW's technical think tank (and wound up being produced in limited quantities because of subsequent demand), laid the groundwork for the Z3. Although it was in some ways impractical—with a composite floorpan and retractable doors—it liberally employed 3-Series mechanical components and handled beautifully. Two traits the Z3 has inherited.

—*Kim Reynolds*

BMW 328

BMW 507

BMW Z1

THE DRIVING SEAT

■ BMW Z3

Film star's variety performance

ANDREW FRANKEL
Road test editor

At last, I have driven the BMW Z3. Until now it had eluded my grasp with unusual dexterity, which is annoying enough with any new car. But what infuriated me about the Z3 was that every single person I spoke to who had driven it felt differently about it. Some thought it the most electrifying small sportscar they'd been in, while others opined that it failed to move the game materially on from the ground already masterfully occupied by the likes of the MGF and Alfa Spider. It was, depending on who you spoke to, both thrilling and boring, ugly and attractive, inspired and cynically derivative.

Encouragingly, it looks the part. I had never been that taken with its shape in photographs, or even in its lacklustre cameo appearance in the *Goldeneye* James Bond film. In the metal, though, it manages to appear both small yet packed with purpose. Its profile is pretty, its nose is menacingly appealing – right back to the shark gills it carries on its flanks – while only the rear appears curiously nondescript. The cabin is plainly based on that of the 3-series compact (as is the chassis), but is no worse for that.

If only it went as well as it looked. Perhaps I should not dwell too much on its performance because the German-spec car I drove was fitted with the 115bhp 1.8-litre single-cam four, while the UK version (when it appears in September) will be fitted with a 140bhp 1.9-litre twin-cam.

Even so, the bigger engine will need to be more sparky even than the figures suggest before it will ask the chassis a remotely thought-provoking question. As it is, the 1.8 provides distinctly mediocre performance offset some way by a fine gearbox, albeit with oddly-stacked ratios.

> "Its profile is pretty, its nose menacingly appealing – right back to the shark gills it carries on its flanks"

I should, apparently, count myself lucky to have driven the Z3 in the wet because at least this gave the chassis an opportunity to provide something more than the allegedly-endless grip it serves up in the dry when fitted with optional 225/50 tyres.

In the rain, it handles beautifully, attacking corners with an ideal whiff of understeer on turn-in which flows gently into broad neutrality or, upon provocation, mild and entertaining oversteer away from the apex. Its communication with the driver is not so involving as a Caterham, nor is its handling so precise but, then again, nor is the Z3 that kind of car. What will matter to BMW is how it fares against Barchettas, Spiders and MGFs, and it is up there with the best. Of its ilk, only the old and less capable Mazda MX-5 proves more involving.

All of which is fine as far as it goes but which, you will have figured by now, is not far enough. The Z3 has colossal presence, is substantial of build, mature in its outlook, and equipped with an image and profile that encourages expectations I fear are far beyond those that its pleasing but hardly-astonishing engines will be able to satisfy.

It is emphatically not a delicate device like a Caterham or even an MX-5 (where small fours make great sense), nor should its multitudinous other talents be denied to those who genuinely enjoy serious driving rather than merely giving that impression. Which means, to these eyes at least, undoubtedly good car though this Z3 is, it fails to realise its true potential by lacking the power to live up to the promise.

What the car needs, I would say, is the assistance of a BMW Motorsport straight-six, probably a 315bhp, 3.2-litre engine. That would knock its 0-60mph time from a little under 10s to little more than five seconds, and bump its top speed up from 116mph to an electronically-limited 155mph.

This course of action would give the chassis the chance to showcase its undoubted talents and provide BMW with one of the most desirable roadsters on the face of the earth – a convincing answer to both the forthcoming Mercedes-Benz SLK and Porsche Boxster. Will BMW build such a car? In fact it already has. It was unveiled at this week's Geneva motor show. ■

The Z3 offers 'mild and entertaining oversteer from the apex'. But will a bigger engine allow this roadster to fulfil its full potential?

ROAD TEST

Rule of Three

Katrina Mueller-Jackson and Henry Rasmussen test BMW's European-designed, American-built iteration of the lightweight sports car.

The point here is easy to miss. Sure, the BMW Z3 is a sharp-handling, crowd-pleasing sort of car. Sure, it's got considerably more interior room than its competition-of-one, the Mazda Miata. Sure, it has that tight, solid, all-of-a-piece feeling for which German cars are famous, and sure, it's been the recipient of more pre-digested PR hoo-ha than any event since Norman Schwarzkopf's book tour.

But to the truly observant car enthusiast, the interesting thing about the launch of the Z3 is not nearly so much what the machine is or what it can do: It's what the little BMW Z3 means for the future of the entire sports-car market. Are we sounding high-falutin' here? Maybe, but consider this: The market for 2-seat sports cars has gone right down the john in the last five years, and its collapse has led to two schools of thought. "Today's aging baby boomer," says the first, "is now too old for small cars, and too chastened by the '80s to ostentatiously display his wealth. Ergo, the market for flashy, impractical sports cars is dead." It's mostly the Japanese who espouse this view, and no big surprise: Having poured billions of dollars into high-tech '90s sports cars and losing a bundle when they flopped, what *else* would they think?

But *"Bullocks and double bullocks,"* reply the largely European members of Camp Two. "Buyers didn't walk away from the sports-car market in the '90s, the sports-

22

car market *walked away from buyers."* And this group has a point. The 1990s saw most 2-place Japanese examples—which had become dominant precisely because they *were* affordable alternatives to costlier European brands—move right into the price range formerly reserved for luxury cars and Porsches. In retrospect, it's probably not surprising that their sales fell faster than a Wallenda with inner-ear problems.

To the Europeans' way of thinking, the Japanese simply bet on a burgeoning world economy and lost. In doing so, their theory continues, they also lost the political will to build any *more* 2-place sports cars, lest they get burned again—and here is where an opportunity can be found.

The result of all this guesswork is that today's enthusiasts are watching a fascinating battle unfold. European carmakers think they can fill the market back up with simpler, less costly sports-cars and revitalize a segment the Japanese gave up for dead.

And it's not just sales that are on the line, either. Since World War Two, the sports-car market has generated publicity and good will all out of proportion to its actual volume or technical importance.

Unto the Breach

Already, into this perceived hole have come the English MGF, Italian Fiat Barchetta and now the German Z3. But while the former two cars are honestly within spitting distance of a middle-class buyer's budget, the Z3 is surprisingly close to the same price range that recently sent the once-popular Toyota MR2 and Dodge Stealth into purgatory.

BMW *has* followed Camp Two's ideals to a point. Yes, the Z3 is an honest, genuine and mechanically simple sports car. Everything about it *was* conceived and designed to be affordable, at least within the framework of BMW's price structure and customer base. The only engine available in America, for instance, is a normally aspirated, reliable 1.9-liter, 140-horse version of their smooth and willing DOHC Four. The platform is that of the well-amortized 3-series sedan, not a unique structure

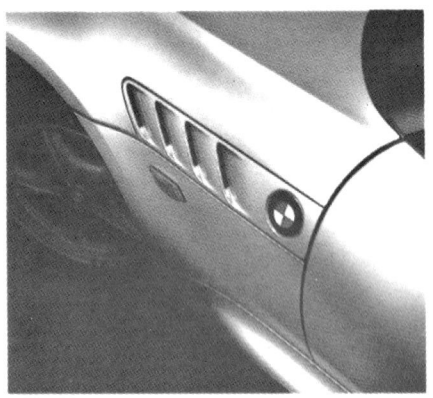

Where the Miata's handling is a whole lot of fun that builds up to fairly low limits, the BMW's is fun, accurate and inspired on its way to very <u>high</u> limits.

developed for just this one car. Suspension is by the current 3-series' MacPherson struts up front and the previous generation's semi-trailing arm setup out back, again to keep things affordable. The roof is operated manually (for now, though that will change later), and standard equipment includes convincing leatherette on the seating surfaces, not genuine dead cow.

The Z3 starts at $28,750, which brings ABS, power seats, mirrors and windows, a/c, a tape player and leather control surfaces. Add leather seats, traction control and some other common goodies, and you can figure that the average car will cost over 30 grand. And while BMW remains mum on the topic, the surely imminent 6-cylinder model will push that price higher still, probably into Porsche Boxster territory.

Now, let's be absolutely, crystal-clear here: I'm not saying for a moment that the Z3 *isn't worth it*. As a matter of fact it *is* worth it, particularly as its price still undercuts its Japanese rivals by thousands of dollars. The point is simply that somewhere between concept and execution, the Z3 ceased to be a truly downmarket, high-volume sportster and became a fairly serious, exclusive proposition.

The Europeans may indeed be coming back into the market, but we're learning that the Z3, Mercedes R-class and Porsche Boxster are not going to be nearly as cheap as we'd hoped: Ten years from now, we may say they correctly identified the problem but come up with the wrong answers. If so, it will be a tragedy of epic proportions for the entire segment.

The Point Being...

Enough crystal-balling for now; what about the matter at hand? Well, the BMW badge alone would sell a ton of Z3s no matter *what* the car cost or how it was built, and with that fact in mind, this roadster seems like a genuine bargain.

Comparisons with the Miata are inevitable, and in most areas the German-designed, South Carolina-built car comes out way ahead. First and foremost, the materials and execution of the BMW are beyond comparison. Where the cheaper, lighter Mazda makes no pretense of solidity or glamour inside, BMW's typically rich, hard plastics and leathers simply exude quality and strength.

The cockpit also feels considerably roomier (as well it should, considering the Z3's 7-inch longer wheelbase). Everything inside will be familiar to BMW faithful, including the firm seats and clear, legible instrument panel. Surprisingly, the big steering wheel is not adjustable, though well-enough placed.

Fire up the engine and for a moment the Miata moves into the lead. The last time we checked, the Japanese car's 1.8-liter engine would move it from 0-60 nearly a half-second faster than our best 8.8-second run in the BMW. The Miata engine also makes much better noises; at best, the German emits a flat, bumbly *whoosh* from the tailpipe which steadfastly refuses to excite. Toss the car into a tight, greasy corner, however, and you immediately see the point of the BMW's bigger window sticker. Where the Miata's handling is a whole lot of fun that builds up to fairly low limits, the BMW's is fun, accurate and inspired on its way to very *high* limits.

At one point in my travels I grossly underestimated the treachery of a slippery, descending S-bend. The resulting combination of bad judgement and a bad roadway would have tossed most any other car—the Miata particularly—over the side, and my own stupid butt right along with it. Not the Z3: Amazing reserves of grip and poise simply planted the car to the pavement, and it obeyed my flailing arms despite all the laws of physics. While the Miata feels like a good old-fashioned sports car, the Z3 comes off as an a modern racetrack refugee. Interestingly, this ability is combined with a smooth and unruffled ride that utterly belies its small size. The Miata could certainly learn a thing or two here.

Turn-in is accurate and crisp and, while arguably antiquated, its healthy serving of traditional BMW drop-throttle oversteer means two tons of fun for the adventurous. Under an extra-heavy foot, the Z3 moves far beyond fun and into the realm of big-league sports and touring machines. A lot of this competence comes down to its extra-rigid structure, well-chosen rubber, standard limited-slip diff and accurate controls. The latter are particularly satisfying, with the Z3's shifter and pedal efforts proving best-in-class. "Yes," I kept thinking behind the wheel, "this is exactly how a convertible sports car *should* feel."

On the outside, cynics may say the most inspiring thing about this car is its badge. I do approve mightily of the Z3's overall proportions—particularly the long-hood/short-deck element that results from the 3-series' 6-banger engine compartment—but many of its details are less to my liking. I'm not crazy about the bulging nose, the scalloped 3rd brakelight or the vertical sidevents *a la*

Z3s will be exported worldwide from BMW's new factory in Spartansburg, South Carolina, the only one building the convertible. Foreign deliveries start first, and these cars offer cloth seats, dealer accessories like luggage racks and a cheaper 1.8-liter, 115-bhp engine not available in America.

SPECIFICATIONS

1996 BMW Z3

General
Vehicle type: front-engine RWD convertible
Structure: steel unibody
Market as tested: United States
MSRP: $28,750
Airbag: std., driver and passenger

Engine
Type: longitudinal L4 with iron block and aluminum head
Displacement (cc): 1895
Compression ratio: 10.0:1
Power (bhp): 138 @ 6000 rpm
Torque (lbs.-ft.): 133 @ 4300 rpm
Intake system: MPFI
Valvetrain: two overhead cams, four valves per cylinder
Transmission type: 5-speed manual

Dimensions
Curb weight (lbs.): 2600
Wheelbase (in.): 96.3
Length (in.): 158.5
Width (in.): 66.6
Track, f/r (in.): 55.6/56.2

Suspension, brakes, steering
Suspension, front: MacPherson struts with control arms, coil springs and antiroll bar
Suspension, rear: semi-trailing arms with coil springs and antiroll bar
Steering type: rack and pinion, power assisted
Wheels, f&r (in.): 16x7
Tires, f&r: 225/50ZR16 (Michelin Pilot HX)
Brakes, f&r: 11.3-inch disc/10.7-inch disc
ABS: std.

Performance
0-60 mph (sec.): 8.8
Top speed (mph): 116 (electronically limited)
EPA est. fuel economy: 23/32 mpg

BMW 507. And while public reaction during my drive was overwhelmingly positive, I had to keep wondering what would have happened if I'd pried off those roundels and glued Pontiac badges in their place. What seems quirky and advanced on an autobahn master often seems unfathomable when it comes from Woodward Avenue.

Still, that comparison is a little disingenuous. Pontiac *isn't* BMW, and without the latter firm's longstanding reputation, the idea of building a $30,000, 4-cylinder, 2-place roadster would never have made it out of the styling room. BMW knows that it's not only offering a simple, basic sports car, here. It's offering a simple, basic *BMW* sports car, and that's something a lot of people will line up to pay for.

They won't be disappointed. ●

BMW Z3 vs MGF

POM & JERRY

PETER ROBINSON plays cat and mouse with two roadsters, products of once deadly enemies and philosophically poles apart, but now in the care a single parent

> One is delicate and civilised, seeking elegant engineering solutions, while the other is

Step out of an MGF and into a BMW Z3 and the contrast is vivid. The kid next door could pick the differences. One appears delicate and civilised, seeking elegant engineering solutions. The other is aggressive, macho, vaguely forced and freely retro in a pragmatic approach to both design and mechanicals.

Now that BMW owns Rover, and therefore MG, they come from the same stable – not that their creators knew this when they began the design process. But they are to be priced as direct rivals and aimed at the same customers.

As approaches to building a two seat ragtop they could hardly be more different, for the contrast extends way beyond technical layouts and visuals. Both impact instantly on the way they drive.

The great irony is that it's BMW which follows a tradition, largely created by MG in the '20s, of putting mass-production mechanicals into a roadster body.

In this case the mechanicals belong to the 318i Compact, complete with the old style semi-trailing rear suspension (not the superior multi-link axle of other 3 Series) and BMW's recently enlarged (to 1.9 litres) 103kW 16v dohc four. These and a mass of other familiar BMW bits are assembled in Spartanburg, South Carolina, and on sale here soon at about $62,500.

MG's approach is to ignore Rover's front-drive hatches in the (did you say correct?) belief that a proper sports car is pushed from the rear wheels. For the MG, Rover's advanced variable valve timing VVC engine, perfectly suited to a sports car, sits transversely between the rear wheels in a mid-engine layout. The MG's 7200rpm redline is a technical leap of 900rpm beyond the BMW's limit and, as you will see, shouldn't be discounted in the contribution it makes to the driving appeal of the car.

Both BMW and MG offer less powerful engines than those fitted here. BMW's is the 85kW 8v 1.8 litre from the 318i. BMW Australia knows it's too weak to be taken seriously and it won't be sold here. Rover's 88kW 1.8 K-series gives a better account of itself, yet the VVC is demonstrably superior. A marque making a comeback, to both sports cars and Australia, after so long deserves the best possible powerplant.

If Rover Australia gets its way, the MGF will go on sale here early next year at a price that should – on the basis of the Pom numbers – undercut the BMW by a couple of grand. That's the theory.

Six months after driving the Z3 in America, nothing has changed my view that while the styling has real presence – I lost count of the number of times I was asked "Is that the

ggressive, macho, vaguely forced, freely retro, pragmatic

James Bond BMW?" – and is handsome from some angles, its proportions are too lopsided for it to be beautiful.

Some love the imposing bonnet and those boldly macho flared wheel arches, but they promise more than the car delivers and you can't help but feel they exist only because BMW plans far more powerful (and longer) six-cylinder versions of the Z3.

A power-to-weight ratio of 11.4kW/kg is at odds with the messages delivered by the Z3's provocative bonnet, diminutive tail and therefore dynamic silhouette. Still, the level of recognition is unbelievable. Given the relatively affordable price (especially in the USA), the Z3 is surely destined to become the most popular of contemporary sports cars – after the MX-5, of course.

The MGF is altogether more feminine in its styling. Softer and far less imposing, more pretty than handsome. The proportions, especially in profile, suggest the Z3 is a much bigger car, but the difference in length is actually only 112mm, and the MGF is 88mm wider.

The British car looks far smaller, more dainty, and more balanced. Yet its 10.4kW/kg power-to-weight ratio promises superior acceleration.

Nor do the interiors contradict the perception that the Z3 is a Compact wearing a roadster body-suit, while the MGF is rather more special, even if it's not quite as well made. Yes, there's more room in the Z3, yet, because much of the furniture is familiar and the driving position uncompromised, there is nothing to distinguish it from any other open BMW.

Electric seat cushion height (but not tilt) adjustment means you can sit low in the Z3, with legs outstretched and no thigh support, or high and comfortable. The MGF feels somewhat like a Honda NSX, with the steering wheel low down between the thighs, partly because the seat cushion is high and the body sides deep. Because the driving position is far more reclined and the windscreen is out of reach, the MG creates the perception it's a baby supercar.

While the view from the Z3 is all bonnet and wheel arch extensions, even tall drivers can't see the steep nose of the MGF.

Neither has brilliant seats when it comes to providing lateral support, though in a straight line they are both comfortable enough; the Z3's instruments – limited to the standard BMW offering – are easier to read than the dials of the Brit. The only left-over from Rover's Honda days are two cheap Civic steering column stalks.

Both have excellent, taut hoods that generate only modest wind noise to around 160km/h. The BMW's roof is easier and quicker to raise, and lower than the more fiddly MGF whose tonneau cover can be a real pain. MG offered a steel hardtop from day one, while BMW's should be available by the time the Z3 goes on sale in Australia.

The Z3's boot is more practical though the space is similar. The MG needs a remote release or at least an opening button.

But you want to know how they drive, these most modern of sporting two-seaters.

On paper, the two appear very closely matched. The recently enlarged BMW dohc four pumps out the same 103kW as the original 1.8 at 6000rpm. The MGF's marginally smaller dohc in-line four develops 107kW at 7000rpm and 174Nm at 4500rpm, using its variable valve timing to good effect in producing a flatter torque curve. The difference in

> Accept the Z3 as Germany's
> disappointed, but the MGF i

overall gearing is minimal but, combined with the peak power rpm, offers an accurate indication as to the character of the two rivals.

Yes, BMW's 1.9 litre engine is smoother and more refined than the old 1.8 but it lacks the sparkle you expect from a sports car and even when it hits its stride at 4500rpm, doesn't sound especially enthusiastic, mostly because of a tinny exhaust note. Just as well it has a quicker gearchange than the MG, for every fraction counts.

There's an urgency about the MGF that's missing from the Z3. It's helped by a 54kg weight advantage, of course, but its engine is just that much more alive and willing, offering a decent growl when it's trying while being just as refined as the BMW while cruising. A long gear lever – both could be shortened to good effect – and slower change that also baulked badly going into third and reverse (I'm assured this is a one-off peculiar to the test car) couldn't spoil the MGF.

The British roadster has the legs of the German in virtually every department and dispatches the standing 400m almost half a second faster.

A time of 15.9sec is terrific for a small 1.8 litre roadster and ensures that it will be treated as a serious sports car.

Quicker through the gears and in fourth and fifth, the MGF always feels more spirited, more eager and, because the engine loves to rev right out to the 7300rpm redline, it is both more responsive and dynamic than the Z3's 1.9. Accept the BMW as a German MX-5 and you won't be disappointed, but the MGF is something rather different and less compromised.

On paper these two deliver exceptional steady-speed fuel consumption, but the reality isn't quite as impressive, the MGF particularly using more lead free if you take advantage of all those revs. The fuel tanks in both are too small to offer a decent touring range.

Both ride brilliantly. Those who reckon real sports cars should only come with the bone-shaking ride of traditional British roadsters will be deeply disappointed at just how comfortable these contemporary interpretations have turned out.

Exceptional body rigidity helps enormously, of course, but the MGF also has the double wishbone suspension and Hydragas springs, while the BMW relies on a more conventional, but equally well sorted, set up. The MGF has all the compliance you expect, without the float and bounce that once went with a system that is a distant, evolutionary development of the hydraulic suspension used on the Morris 1100 way back in the '60s.

The MGF shrugs off bumps, absorbs potholes and treats rough roads with contempt, yet the BMW's low speed ride is even better.

Both have assisted rack and pinion steering, the MGF

sing an unusual electric power assistance. Lighter and ss precise than the Z3, the 1G's steering weighs up rogressively in corners but ever feels as natural or quick its rival. Come from the 1G to the Z3 and you dart ound, turning in too quickly, ough the difference between 9 and 3.1 turns lock to lock ems marginal and should also e related to the BMW's ghter turning circle. The Z3 mply has more direct and eatier steering that creates a ositive first impression.

Both suspensions have en set up to provide mild d reassuring initial ndersteer, in the modern shion. Up to eight tenths, it's e BMW that feels the more njoyable and confidence-spiring, mostly because of e steering's more naturally volving messages. Push eyond that level and you uickly discover the MGF tains its body composure and lance, stays the more neutral d will eventually move to an sy-going oversteer.

Ultimately, a combination the right tyres, that effortless ngine and a wonderful poise d true adjustability of chassis eans the driver of the MGF els more an integral part of e experience of the car.

The Z3, competent and oroughly predictably BMW its behaviour, will eventually ift into oversteer, especially hen the road is wet, but it kes an almost brutal effort d hasn't the intrinsical ythm of the MG. Maybe this s everything to do with the

grip of the massive 225/50R16 Michelins fitted to the test Z3 in place of the standard 205s which surely offer a better mix of communication and adhesion. Deliberately back-off at the corner's apex to measure the change in attitude and there's a momentary pause before the Z3 reacts.

Body roll intrudes and bumps that the MGF takes in its stride induce a sideways hop from the rear suspension.

Both have excellent brakes. Those on the BMW are probably superior under duress though initially feel a little dead, while the MG shows that its suspension isn't perfect by dipping the nose.

I've no doubt the Z3 will be produced in numbers more vast than Rover will dream of for the MGF. The cachet of the spinning propeller, the instinct that it's going to be better built and slightly easier to live with day to day, the certain knowledge that everybody is going to know and recognise the BMW is one powerful marketing tool.

Yet, despite steering that should be better weighted and a nasty gearchange. I'd rather drive the MG through my favourite stretch of the twisties. It delivers a more special drive and is more of an event than any journey taken in the Z3.

Of course, the fact that they are so different gives rival car makers no joy, for it means that BMW (especially when the mouth-watering M-roadster arrives), as proprietor of both the Z3 and MGF, has the class covered.

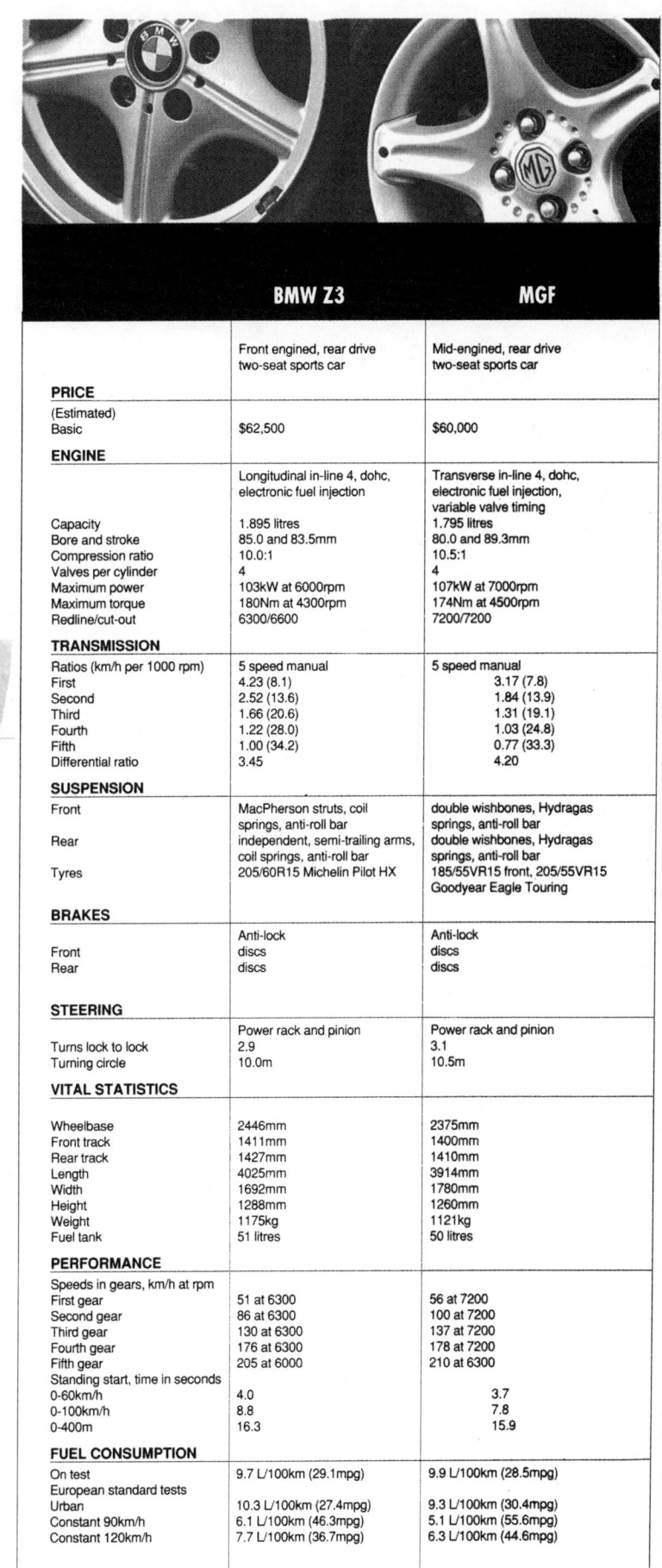

> X-5 and you won't be omething rather different and less compromised

	BMW Z3	MGF
	Front engined, rear drive two-seat sports car	Mid-engined, rear drive two-seat sports car
PRICE		
(Estimated) Basic	$62,500	$60,000
ENGINE		
	Longitudinal in-line 4, dohc, electronic fuel injection	Transverse in-line 4, dohc, electronic fuel injection, variable valve timing
Capacity	1.895 litres	1.795 litres
Bore and stroke	85.0 and 83.5mm	80.0 and 89.3mm
Compression ratio	10.0:1	10.5:1
Valves per cylinder	4	4
Maximum power	103kW at 6000rpm	107kW at 7000rpm
Maximum torque	180Nm at 4300rpm	174Nm at 4500rpm
Redline/cut-out	6300/6600	7200/7200
TRANSMISSION		
Ratios (km/h per 1000 rpm)	5 speed manual	5 speed manual
First	4.23 (8.1)	3.17 (7.8)
Second	2.52 (13.6)	1.84 (13.9)
Third	1.66 (20.6)	1.31 (19.1)
Fourth	1.22 (28.0)	1.03 (24.8)
Fifth	1.00 (34.2)	0.77 (33.3)
Differential ratio	3.45	4.20
SUSPENSION		
Front	MacPherson struts, coil springs, anti-roll bar	double wishbones, Hydragas springs, anti-roll bar
Rear	independent, semi-trailing arms, coil springs, anti-roll bar	double wishbones, Hydragas springs, anti-roll bar
Tyres	205/60R15 Michelin Pilot HX	185/55VR15 front, 205/55VR15 Goodyear Eagle Touring
BRAKES		
	Anti-lock	Anti-lock
Front	discs	discs
Rear	discs	discs
STEERING		
	Power rack and pinion	Power rack and pinion
Turns lock to lock	2.9	3.1
Turning circle	10.0m	10.5m
VITAL STATISTICS		
Wheelbase	2446mm	2375mm
Front track	1411mm	1400mm
Rear track	1427mm	1410mm
Length	4025mm	3914mm
Width	1692mm	1780mm
Height	1288mm	1260mm
Weight	1175kg	1121kg
Fuel tank	51 litres	50 litres
PERFORMANCE		
Speeds in gears, km/h at rpm		
First gear	51 at 6300	56 at 7200
Second gear	86 at 6300	100 at 7200
Third gear	130 at 6300	137 at 7200
Fourth gear	176 at 6300	178 at 7200
Fifth gear	205 at 6000	210 at 6300
Standing start, time in seconds		
0-60km/h	4.0	3.7
0-100km/h	8.8	7.8
0-400m	16.3	15.9
FUEL CONSUMPTION		
On test	9.7 L/100km (29.1mpg)	9.9 L/100km (28.5mpg)
European standard tests		
Urban	10.3 L/100km (27.4mpg)	9.3 L/100km (30.4mpg)
Constant 90km/h	6.1 L/100km (46.3mpg)	5.1 L/100km (55.6mpg)
Constant 120km/h	7.7 L/100km (36.7mpg)	6.3 L/100km (44.6mpg)

BMW Z3

Everything you want in a regular Z3, and more

BY MATT STONE
PHOTOS BY THE AUTHOR

You've heard them all: those clever clichés about bigger engines being better engines. "Ain't no substitute for cubic inches." "There's no such thing as too much horsepower." "Speed costs money—how fast do you want to spend?" BMW has taken its already fabulous Z3 Roadster, and given it an American style—literally and figuratively—engine swap, to come up with the new Z3 2.8. The result is the opposite of Lite beer: Everything you want in a regular Z3, and more.

In last year's edition of *Sports & GT Cars* we named the Z3 "Sports Car of The Year," and with good reason. It's a clearly focused, purely conceived sports car that's uniquely styled, beautifully built and a ball to drive. We love the Z3's viceless handling, willing drivetrain, smart ergonomics and more-than-comfy ride balance. Apparently so do you, as nearly every one of the 250 or so Z3s that roll out of BMW's Spartanburg, South Carolina, plant every day are spoken for before the paint dries.

Some critics say the Z3 is underpowered. We say not true. Driven as a sports car by a driver who uses the 1.9-liter dohc 4-cylinder's power band, understands that heel-and-toe is not a version of the Texas Two-Step, and gets all slurpy over a perfect downshift, the Z3 delivers sublime satisfaction. The issue is that the chassis is so solid and the handling so well balanced, it just begs the question: "What would this be like with more beans?"

BMW's stock-in-trade since the '60s has been its delicious straight-six powerplants. The 2.8-liter (2793-cc) version that has found its way into the Z3's snout is a part of the dohc "small" six engine family, though it differs from the one found in the 528i sedan in several ways. It's the first U.S. usage of the aluminum-block version, lighter than the iron-block unit by 53 lb. It's rated at 189 bhp and 203 lb-ft of torque, down a bit from the sedan version; the difference lies in the Z3's single exhaust system. Still, that's a 52-bhp boost over a 4-cylinder Z3, representing a 38 percent increase in the bean department.

Everything else is what you've come to expect: double overhead cams, direct-port fuel injection, 4-valves-per-cylinder, as well as BMW's own engine management and VANOS variable cam-adjustment systems. We might as well get to the big question right now: How much faster is it? Lots. Last year we tested the 4-cylinder Z3 to a 0–60 time of 8.1 seconds. BMW says the 2.8 covers the same ground in 6.3. Though we've not been able to run our standard battery of tests just yet, a couple of stopwatch runs on the less-than-dragstrip-flat back roads of Madeira (a Fantasy Island-like place off the coast of Portugal where BMW previewed the new car) indicate that number is plausible. Backing the silky six is BMW's excellent 5-speed manual transmission, and an electronically controlled, multi-shift-mode 4-speed automatic is available as an option ($975).

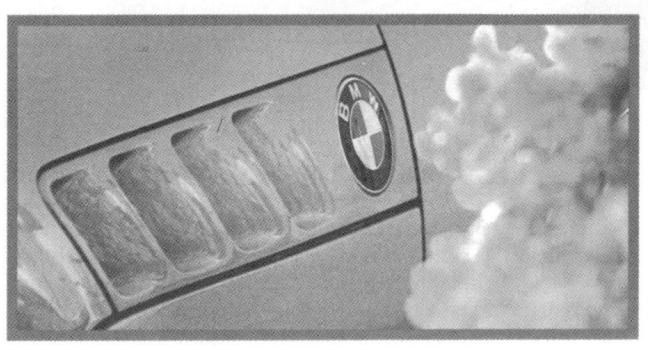

It would be un-BMW-like to just drop in the big motor and run, so the chassis, interior and option list also get a workover. Front track is widened by a scant 2 mm, but rear track goes up 2.5 in. to 58.8 in., all in the name of enhanced stability. A limited-slip diff is standard, and working in concert with BMW's Automatic Stability and Traction Control (ASC+T), they endeavor to keep the Z3 planted no matter how ham-fisted the driver gets with it. For those wanting to dial a little dirt-tracking into the handling mix, BMW courteously provides an off switch. Shock absorbers, anti-roll bars and springs have all been recalibrated to handle the slight increase in weight, a mere 90 lb says BMW, and the weight balance is spot on 50/50.

Appearance-wise, the differences are subtle, but the steroid injection does show. The front air intake is larger and there's now a dual-tip exhaust, but the 2.8 version has really grown in the hips. The rear fenders are suggestively flared by 3.4 in., and when the car is viewed from above, it smacks of a 7/8ths-scale Viper. Besides casting a more aggressive look, the wider bustle serves to cover up the optional wider tires.

While the four-cylinder's 16x7 alloy wheels and 225/50ZR-16 tires are also standard on the 2.8, BMW now offers an optional 17-in.-wheel package that calls for 7.5 inchers on the front, 8.5-in. units on the back, with 225/50ZR-17 and 245/40ZR-17s respectively. The backs are about the same size as found on a Lotus Esprit S4, a much larger car. Both the standard and optional wheels are of a new rounded-spoke design, to cast another subtle differentiation between the 2.8 and its little bro. But talk about subtle, there is no exterior badging to hint at the increased horsepower contained therein. Perhaps all the better for teaching unsuspecting Miatas a lesson. ...

We've liked the Z's interior's smart layout, easy-to-use controls and functional look from the outset, but some felt it lacked a bit of pizzazz. Well now you can order up a little bit more, or a whole bunch more. Leather upholstery is standard on the 2.8, though our *arktissilber* tester had the optional two-tone treatment, in mini-skirt red and black. Eyepopping, but not exactly our taste. A mere $150 buys the Chromeline package, which adds chrome bezels to the gauges (nice), chrome interior door handles and handbrake release (OK), and a chrome headlight switch that looks as if it were ripped out of a '66 Plymouth Belvedere (tacky). BMW does a fine job with its wood trim, in this case a well-burled walnut, and it adds warmth and class to the cabin. Also available (by April, 1997) is a power-operated soft top, though the manual unit is so easy to raise and lower it almost makes the power setup redundant. The new folding windscreen reduces noise and buffeting in the cockpit when cruising topless.

One complaint we've always had about the original Z3 is its less-than-inspiring exhaust note. The 2.8 replaces the 4-banger's metallic fizz with the deep, snarly, high-quality tone that somehow only seems to come from a

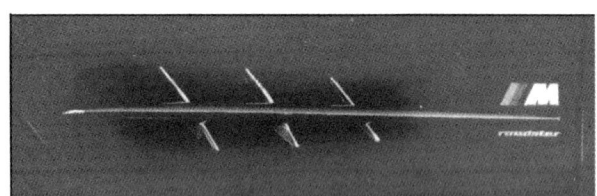

BMW Z3 M
Not available here...yet

well-tuned straight six. It burbles softly upon engine braking, and enhances the car's performance character immensely. I would even enjoy 10-15 percent more of it, but the aftermarket will surely have cat-back exhaust systems available soon enough, to meet the need of every ear.

American drivers, bottle fed on V-8 engines nearly from birth, will appreciate the 2.8's generous torque curve. It pulls solidly from a low 1500 rpm, and peak torque is reached at 3950. It flattens slightly just over 5000, and drops a bit more as it approaches the 6500-rpm redline, but overall the powerband cuts a smooth, wide path. Working the engine up and down between 2000 and 5000 is Happyland, and it never gets harsh or buzzy. Just another of BMW's inline-6s doing its thing.

The gear ratios of the 5-speed stick seem ideally matched to the engine. They allow the Z to come out hard, and you can even light the 245-section tires at will, yet 70-mph cruising is almost lazy. Clutch effort is light, yet take up is smooth with zero chatter. The gearshift falls easily to hand and the shift action is crisp, though it still lacks that sweet snickability of the Mazda Miata's shifter. We did not sample the automatic, but suspect that it makes a good mate for the new powerplant.

Adding cylinders, equipment and a few pounds often come at the cost of reduced handling prowess, or at least some steering response, but we could find no detriment whatsoever. As noted, weight distribution goes from 52/48 to an even 50/50, and the Z3's already superb balance is unaffected. The power increase more than offsets the minuscule weight gain, and enhances the drive since you can now provoke a bit of oversteer with your right foot when desired. But don't worry about any sudden end-swapping on this relatively short-wheelbase roadster; the 17-in. wheel/tire package really keeps both front and back-end firmly in contact with the Tarmac.

The road quality on Madeira varies from acceptably smooth 2-lane pavement, to single-lane oxcart trails with more than their share of chuckholes and marginal patching, yet the handling balance remained unfettered. The addition of +1 rolling stock often comes at the expense of ride quality, but not in this case. Our 2.8 rode every bit as well—and as quietly—as a car with the base 16-in. combo, though we wonder if this will be the case after the low-profile tires have 15,000 or 20,000 miles on them. Ride suppleness is almost luxurious without being ponderous.

Turn-in provided by the speed-sensitive power steering feels a bit crisper with no increase in steering effort. It has all the brakes it can use, hauling the Z down hard with outstanding

---

Just like your favorite salsa, Z3s will soon be offered in Mild, Hot and 3-Alarm Fire versions. BMW's M Division has been casting its spell over the Z, and the result is the M roadster, which will be available for outside-U.S. delivery sometime this spring. Why not us!?! Hang on a minute. ...

The Z3 M gets the Euro-version M3's 3.2-liter six, good for 321 bhp. It should clip at least another second of the 2.8's 0-60 time of 6.3 sec. A more-aggressively styled front fascia with wire mesh covering the air intakes, huskier fender flares, wider 17-in. wheels, M badging and twin dual exhaust pipes give the M roadster that testosterone-inspired look. Most unique, and attractive, are revised air vents in the front fenders that look for all the world as if they came off a 507. And not by accident.

The cockpit will get grippier sport seats, an M-Sport steering wheel, more chrome trim, individual roll hoops behind the seats—quite the rage these days—and a trio of ancillary gauges in the console, which would be a positive addition to the 2.8 as well (BMW, are you listening?). Figure that the M will also get a makeover at the suspension spa. Pricing has not yet been announced.

What makes the M roadster even more unique is that it's the first BMW M product that is built entirely outside of Munich. Yes, they will be built in Spartanburg, alongside all the rest of the world's Z3s. Word around Madeira was that we American's will get our own M roadster soon enough, though expect a U.S. spec, 240-bhp M3 engine to replace the export 321-bhp unit by the time the car is offered here. Oh well, I guess we can still get burned by a 2-alarm fire for now.—*Matt Stone*

modulation. We always appreciate BMW's anti-dive geometry, still among the best around. The interior packaging helps too: there's leg room aplenty, the seats grip solidly and comfortably, and there's more than adequate head room with the top up. Journalists are trained to describe things in the specific and not the amorphous, but we have to say it; it just feels like a BMW.

There's little not to like here. As noted, the yowzer two-tone-and-chrome treatment is not for everyone, but thankfully both are options, allowing the buyer to dial in as much or as little tart as they desire. Though the interior is well finished as a whole, both the plastic glovebox on the console and the lockable storage compartments behind the seats have a substandard look and feel. The locks give the impression they could be pried open with a strand of overcooked linguini, and both have been the source of rattles in Z3s we've driven.

The Z3 2.8 accomplishes several things for BMW. Whether they like it or not, comparisons with the Miata have been inevitable. But, Mazda does not offer anything beefier than the standard 1.8-liter four, so the 2.8 version moves the Z3 into a different class. Buyers who would not even consider a

Z3 (or a Miata) because of 4-cylinder insecurity will take to the 2.8's power and acceleration in a heartbeat. It also gives the company a weapon for a market segment that's about to become very crowded: the Porsche Boxster and Mercedes-Benz SLK are coming hot on the 2.8's heels. Though there are differences in their design, personality, performance characteristics and target buyer, it still adds up to three new somewhat comparably conceived roadsters hitting the block at the same time—and aiming for some of the same dollars. Clearly BMW anticipated this: with a well-equipped base price of $35,900, it will undercut the new Porsche by at least $4000. The reach to the SLK will likely be greater still. The plant's production mix will be split approximately 50/50 between 4 and 6-cylinder versions at the outset, and all Z3s continue to be built in Spartanburg for worldwide distribution.

It would be easy to expect a compromise in the process: get some things, give up some things. But the 6-cylinder engine and other enhancements that make a Z3 into a Z3 2.8 are all genuine improvements and do nothing to diminish the package's basic goodness. If speed costs money, then it's value received for dollars invested.

BMW

Z3 2.8

PRICE
List price, all POE	$35,900
Price as tested	est $38,900

Price as tested includes std equip. (dual airbags, a/c, speed-sensitive pwr steering, ABS, leather upholstery, cruise control, central locking), options on test car consist of 17-in. alloy wheel/tire package ($1000), metallic paint ($475), 2-tone interior trim (est $400), Chromeline interior trim ($150), luxury tax (est $405), dest charge ($570).

ENGINE
Type	dohc 24-valve inline-6
Displacement	2793 cc
Bore x stroke	84.0 x 84.0 mm
Compression ratio	10.2:1
Horsepower, SAE net	189 bhp @ 5300 rpm
Torque	203 lb-ft @ 3950 rpm
Maximum engine speed	6500 rpm
Fuel injection	elect. sequential port
Fuel requirement	premium unleaded

GENERAL DATA
Curb weight	2778 lb
Weight distribution, f/r, %	50/50
Wheelbase	96.3 in.
Track, f/r	55.6/58.8 in.
Length	158.5 in.
Width	68.5 in.
Height	50.7 in.
Trunk space	6.4 cu ft

CHASSIS & BODY
Layout	front engine/rear drive
Body/frame	unit steel
Brakes, f/r	11.3-in. vented discs/11.0-in. discs, vacuum assist, ABS
Wheels	cast-alloy, 17x7.5 f, 17x8.5 r
Tires	Michelin Pilot SX 225/50ZR-17 f, 245/40ZR-17 r
Steering	rack & pinion, speed-sensitive power assist
Turns, lock to lock	2.7
Suspension, f/r	MacPherson struts, lower L-arms, coil springs, gas-charged twin-tube shock absorbers, anti-roll bar/semi-trailing arms, coil springs, gas-charged twin-tube shocks, anti-roll bar.

DRIVETRAIN
Transmission		5-sp manual
Gear	Ratio	Overal Ratio
1st	4.20:1	13.23:1
2nd	2.49:1	7.84:1
3rd	1.66:1	5.23:1
4th	1.24:1	3.91:1
5th	1.00:1	3.15:1
Final-drive ratio		3.15:1

PERFORMANCE DATA
0-60 mph	6.3*
Lateral accel (200-ft skidpad)	0.93g*

FUEL ECONOMY
Normal driving	est 23 mpg
Fuel economy (EPA city/hwy)	19/27 mpg
Fuel capacity	13.5 gal.

*Manufacturer's estimate
est estimated
na means information is not available

SHOWD

OWN!

Does BMW, Mercedes, or Porsche build the most desirable roadster?

BY MARK GILLIES

You know that life for car enthusiasts is good when pillars of automotive rectitude like BMW, Mercedes-Benz, and Porsche start building small sports cars that have the whiff of affordability about them. This is the first time since the Fifties that we've had a choice of beguiling roadsters from the three most prestigious German automakers.

But life has changed since the Fifties. In those days, cars like the Porsche 356 Speedster stood out like ocean liners in a flat sea full of rowboats. The difference between good and bad cars was vast. Nowadays, virtually every automaker builds cars that combine excellent performance, roadholding, and handling with safety and economy. When someone builds a sports car, it's a given that it will look great and go fast. But we want more than mundane, go-fast goodness—what we're looking for is automotive greatness.

That's what we're here to find out. Which of these three has the most character? Which connects best with you, the driver and passenger? Which of these three is the "It Car"?

All three will impress your local valet parkers, and not just because of the upmarket badges adorning their noses. Arguably, the Boxster is the cutest of the three and the SLK the prettiest, but they're all good-looking, desirable roadsters. The Z3 2.8 certainly looks better than its four-cylinder sibling, thanks to a more aggressive front spoiler and a wider rear track that fills out raunchier arches, but they can't disguise the slight weakness around the car's hind quarters. The Boxster is terrific, but it's difficult to forget that the original show car of 1993 was such a knockout.

Inside, the funkiest and most stylish is the Mercedes, with its white-faced dials and carbon fiber center console. You either love it or loathe it, whereas the Porsche has clean, attractive shapes—including the coolest door handles on the planet—and great retro detailing, like the three-into-one gauge cluster and the shiny black finish on the minor controls. Alongside these two, the Z3 is pleasant inside, but nothing special: first-rate ergonomics (although entry and exit is difficult for tall people due to narrow door openings), good materials, and pleasant shapes, but the Z3 interior is more like one you'd expect from Mercedes. Happily, you can option it out with jazzier trim.

The BMW is also lacking in the top-up department. That is, it's the only one here that doesn't have an electric top mechanism. Manual only, until spring, when a power option becomes available. The other two, by contrast, have power tops that could have come out of *Mission Impossible:* The Porsche's takes twelve seconds to do its stuff, while the clamshell Mercedes metal top takes twenty-five seconds and will keep friends and family amused for hours. That

PHOTOGRAPHY BY BILL CASH

The 189-bhp six-cylinder engine transforms the Z3, making it far more responsive to a driver's demands than the four-cylinder. Z3 2.8 buyers get a good basic specification, with cruise and traction control, leather trim, a wooden gear lever, and wood console facings. Chrome for the light switch and gauges costs extra.

SHOWDOWN!

same clamshell arrangement drastically compromises trunk space when stowed, however. Surprisingly enough, the Porsche wins on practicality because there's simply more room for people and their chattel. All three cars show how far the soft-top science has come in the past twenty years, because it's possible to converse normally at 100 mph inside, all three are snug and draftproof, and all look equally good with the top in place.

Technically, the Boxster promises the most. Where the other two have front-engine, rear-wheel-drive layouts, the Porsche goes with a mid-location for its flat-six, twin-cam engine. It has the most power, with 204 bhp (DIN) compared with the supercharged 2.3-liter four-cylinder SLK's 191 bhp and the straight-six 2.8-liter BMW's 189 bhp. Like the other two, it has four-wheel independent suspension and anti-lock brakes: But whereas the BMW and the Mercedes have vented front discs and solid rear rotors, the Porsche has vented disc brakes all the way around.

Out on the road, you're going to have big fun in all three. The Z3 2.8 addresses the one major flaw with the original: a nice car looking for a real engine. The four-cylinder Z3's 138-bhp unit not only lacks punch, but it makes a whiny four-cylinder noise. The twin-cam six, familiar to 3- and 5-series drivers, not only bumps top speed up nineteen miles per hour to 135 mph and shaves nearly two seconds off the 0-to-60-mph time—down to just 6.3 seconds—it provides the midrange power the car needed. In a straight-line sprint this is the quickest of the three, although its top speed is eight miles per hour down on the Mercedes and fourteen away from the Boxster.

Given the power it deserves, the Z3 2.8 becomes a much more compelling proposition. For one, it sounds lovely. And second, there's now enough power to exploit the inherent goodness of the Z3's chassis. With the four-cylinder engine, the power/grip equation feels like it is biased in favor of grip, a flaw that doesn't, for instance, affect the Mazda Miata. With the six, the extra power can be used to dial out front-end push, and the enhanced throttle adjustability means it's much more satisfying to drive. It feels like a faster, more grown-up Miata, which is what it always should have felt like.

Throw in excellent brakes, a very slick gearshift, a pretty good ride, and fluid steering, and you have one fine sports car. It isn't sophisticated in the way it drives, but it is definitely fun. (Don't be tempted by the seventeen-inch wheel option with 225/45 front and 245/40 rear tires, because the car feels better balanced on the sixteen-inch rims with 225/50 tires all around.)

The SLK feels more serious and sophisticated than the BMW. Less of a sports car, in fact, and more of a pint-sized Grand Tourer. The supercharged four-cylinder engine produces peak torque of 206 pound-feet between 2500 and 4800 rpm, so there's great midrange performance. You expect a supercharged engine to make a good noise, yet this one wheezes low down and rasps high up. It's not unpleasant, but it doesn't get the juices flowing. The engine's effectiveness is partially masked by the five-speed automatic transmission with which Mercedes has saddled U.S. cars: It just doesn't feel right in something with sports car pretensions. Somehow, the ratios don't marry well with the engine's torque curve.

Move down a testing road, and the Mercedes evinces the structural rigidity of a house. It is also less deft than the BMW or the Porsche, with a dead spot in the steering and a tendency to heave its mass onto the outside rear wheel while cornering hard. The freeway ride is supple, but its ability to soak up bumpy secondary roads is compromised. Ultimately, the chassis is safe and secure, and you can force it into having fun, but it's not as easy to balance on the throttle as the BMW Z3.

Most drivers, most of the time, will love the Mercedes. But despite excellent brakes, a good ride, and bags of grip, the SLK isn't a car that delights you down a favorite road.

Surprisingly, the SLK can be flung around, but it takes abuse to relinquish grip from the fat tires. Inside, flat seat cushions and too much lumbar curvature mar what is otherwise a fine driving position. There's plenty of oddment space, and the retro-style, chrome-rimmed dials look terrific.

IN THE
BOXSTER, IT DOESN'T
FEEL AS IF YOU'RE
MASTERING A CAR. THE
SLK AND THE Z3 OPERATE
ON A LOWER LEVEL,
WHERE YOU DRIVE AND
THE CAR FOLLOWS.

SHOWDOWN!

It's not as quick away from the line as the Z3 or the Boxster—0 to 60 mph takes 7.2 seconds, compared with 6.3 for the BMW and 6.7 for the Porsche—and it doesn't react as sharply or as neatly as the other cars.

The Porsche has its flaws too, but it is the best car here. The Boxster is It, period.

The engine has been criticized by some for lacking power, but what it actually lacks is low-end power. The Boxster's flat-six produces its maximum power at 6000 rpm and its peak torque of 177 pound-feet (DIN) at 4500 rpm: Even the BMW makes more torque—203 pound-feet—at lower revs, 3950 rpm. To get the best out of the Boxster, you need to spin the engine into the upper reaches of the tach dial and make plenty of use of the five-speed transmission. (The Tiptronic S five-speed automatic transmission is available, but the self-shifter is far more appealing.)

Using plenty of revs is no chore because the shift is fast and positive, and the engine makes the kind of spine-tingling flat-six noises we have come to expect from a Porsche, though it's more of a wail than the 911's growl. It's an aural treat even at low revs, and pressed hard it sounds stunning. It revs freely and eagerly but feels lethargic if you hang around in high gears at low rpm.

Where this car really scores in this company—make that any company—is in its tactile responses. The steering has a meaty feel, linear response, directness, and communicative feedback that makes it the best here. The brake pedal has the most progressive action, and the awesome brakes are the most powerful of the trio. Pedals and controls feel like they have been honed by people who love and understand driving.

The Boxster is so poised, it is the diplomat of the breed. The chassis is sensitive to your every input and reacts instantly. Unless you're a maniac, you won't be booting the

The production Boxster may not look as great as the show car, but the shape is attractive and will last well. The leather interior costs extra, as do traction and cruise control. The driving position is the most intimate of the three cars, the ergonomics first-rate.

tail out under power, but you will be reveling in a car whose cornering attitude is dictated by subtle changes in the throttle and the steering. There's huge grip and traction is brilliant: That favorite country road becomes an arcade game, a real rush. The ride is the firmest here, but the upside is the best body control of the three. In truth, all three roadsters move across the blacktop better than many sedans.

The Boxster begs the question, Do you need enormous amounts of horsepower to have fun? Well, not here, especially if you opt for the Boxster on its standard 205/55 front and 225/50 rear tires on sixteen-inch rims, rather than over-tire it with the optional seventeen-inch wheels. In the Boxster, it doesn't feel as if you're mastering a car, rather that you're marrying it. The SLK and the Z3 operate on a lower level, where you drive and the car follows.

All three cars will give immense pleasure to their owners. The SLK comes across as a true heir to the SL tradition, exemplified by the 190SL of the Fifties and the 230/250/280SL series of the Sixties. Pretty and stylish, it's a nice roadster that doesn't quite live up to its sports car billing. When we were photographing the three cars around Santa Barbara, the SLK was the one that attracted the ladies, which probably tells you where most of them will end up.

The Z3 2.8 is a real sports car by comparison. Compared with the SLK, it is more enjoyable to drive fast, it's quicker off the line, and it makes the right kinds of noises. (We'd have said the same things about a BMW 507 in the Fifties, too, compared with a 190SL.) It also offers the best value of the three, at a $35,900 base price, with leather, cruise, and traction control standard. If you can't stretch to the Boxster—and comparably equipped, the Porsche weighs in at $43,000—then this is a really good substitute. Is it worth $15,000 more than a Miata? We think so: It's now the sports car it always promised to be.

Which brings us to the Boxster. No, it's not cheap, and it's hard to justify a base price twice that of a Miata—but the Boxster does it. Like the 356 of the Fifties, the Boxster shows that sports car greatness isn't about speed alone. It's a measure of how much fun you can have, how alert the car is to your demands, and how rigorously it has been developed and engineered. Strip away the style, the badges, and the hype, and what you're looking at in the Boxster is a car that redefines the breed. ■

BMW Z3 2.8
Front-engine, rear-wheel-drive roadster
2-passenger, 2-door steel body
Base price $35,900/price as tested $36,470 (+ luxury tax of 8% over $36,000)

POWERTRAIN:
24-valve DOHC 6-in-line, 170 cu in (2793 cc)
Power SAE net 189 bhp @ 5300 rpm
Torque SAE net 203 lb-ft @ 3950 rpm
5-speed manual transmission

CHASSIS:
Independent front and rear suspension
Variable-power-assisted rack-and-pinion steering
Vented front, rear disc brakes
Anti-lock system
225/50ZR-16 Michelin Pilot HX tires

MEASUREMENTS:
Wheelbase 96.3 in
Length x width x height 158.5 x 68.5 x 50.9 in
Curb weight 2778 lb

PERFORMANCE (manufacturer's data):
0–60 mph in 6.3 sec
Top speed 135 mph
EPA city driving 19 mpg

PORSCHE BOXSTER
Mid-engine, rear-wheel-drive roadster
2-passenger, 2-door steel body
Base price $39,980/price as tested (estimated) $43,000 (+ luxury tax of 8% over $36,000)

POWERTRAIN:
24-valve DOHC horizontally opposed 6, 151 cu in (2480 cc)
Power DIN 204 bhp @ 6000 rpm
Torque DIN 177 lb-ft @ 4500 rpm
5-speed manual transmission

CHASSIS:
Independent front and rear suspension
Variable-power-assisted rack-and-pinion steering
Vented front and rear disc brakes
Anti-lock system
205/50ZR-17 front, 255/40ZR-17 rear Pirelli P Zero tires

MEASUREMENTS:
Wheelbase 95.1 in
Length x width x height 169.9 x 70.1 x 50.8 in
Curb weight 2761 lb

PERFORMANCE (manufacturer's data):
0–60 mph in 6.7 sec
Top speed 149 mph
Fuel economy, European cycle 29–36 mpg

MERCEDES-BENZ SLK
Front-engine, rear-wheel-drive roadster
2-passenger, 2-door steel body
Base price (estimated) $39,900/price as tested (estimated) $41,500 (+ luxury tax of 8% over $36,000)

POWERTRAIN:
Supercharged 16-valve DOHC 4-in-line, 140 cu in (2295 cc)
Power SAE net 191 bhp @ 5300 rpm
Torque SAE net 206 lb-ft @ 2500–4800 rpm
5-speed automatic transmission

CHASSIS:
Independent front and rear suspension
Variable-power-assisted recirculating-ball steering
Vented front, rear disc brakes
Anti-lock system
205/55R-16 89V front, 225/50R-16 92V rear Michelin Pilot SX tires

MEASUREMENTS:
Wheelbase 94.5 in
Length x width x height 157.3 x 67.5 x 50.8 in
Curb weight 2922 lb

PERFORMANCE (manufacturer's data):
0–60 mph in 7.2 sec
Top speed 143 mph
Fuel economy, European cycle 29 mpg

Straight six pushes BMW Z3 into the lead

The Z3 roadster has been transformed by BMW's straight six. Andrew Frankel discovered it has the grunt to match its looks

The die-hard traditionalists continue to hold sway at BMW. There can be no variety of sportscar more steeped in history than the open two-seater, and Munich's latest intepretation, the BMW Z3 2.8, remains faithful to the breed.

There's nothing clever or visionary about creating a roadster these days, as everyone's doing it. From a position at the start of the decade where Mazda's MX-5 ruled the roost almost unchallenged, the maker who now does not possess a roadster is, indeed, missing a trick.

It is the Z3's conformity to roadster values that makes it different.

Six cylinders are the making of BMW's Z3, offering the poke its stable-mate lacks

Rivals such as the Alfa Spider or Fiat Barchetta are converted from hatchback platforms and share their front-wheel drive layouts, while the MGF moves the hatchback driveline assembly behind the driver so it now drives the rear wheels. The Porsche Boxster, too, is mid-engined.

The Z3 is concerned with no such trickery. Its large capacity, multi-cylinder engine is under the bonnet, its drive directed to the rear wheels alone. Not that this, in any way, militates against the Z3.

The packaging disadvantages of a longitudinally mounted six-cylinder are inconsequential in a two-seater and add to the car's visual appeal, thanks to the evocative long bonnet.

Indeed, it is this engine which provides the Z3's best argument. BMW knows how to do a straight six better than anyone, and this 2.8-litre version, with its twin chain-driven camshafts and 24-valves, offers 193bhp. This may not sound too exciting, but bear in mind first that it's a largely artificial figure designed as a tax dodge for the German market and, second, it's backed up by a fat 202lb ft of torque.

The result is the car that the Z3 has always threatened but, with four-cylinder power, ultimately failed to be. It now has the performance to carry off its sharkish looks, hitting 60mph in under 7sec and going on to a noisy 135mph. Variable valve timing ensures solid response at all points in the powerband, while the exhaust note, muted at idle, rises to a crescendo that stands out in stark comparison to the annoying drone of the Mercedes SLK. But you do wish that the show would go on a little longer before reaching for the next gear; as peak power arrives at a mere 5300rpm, while the fuel feed is cut at 6400rpm.

Despite using the rear suspension from the previous 3-series, the oft-recalled twitchiness it was alleged to induce has been exorcised. Wet or dry, under extreme provocation, it will not misbehave. Sure, it's not entirely foolproof but, even if you turn off the standard traction control, you'll not shake the tail loose unless you're trying hard in the wet.

What remains is a chassis of consummate ability which needs a little more character to really sparkle. Point to point speed is all very well, but if it's not accompanied by the kind of balance and adjustability that some drivers crave, its appeal will fall far short of being all-encompassing.

In most other respects, the Z3 makes a fine roadster. The hood mechanism is both simple and convenient and while it will not fool you into thinking it's a coupe with the hood up, like the SLK, it is still refined on long journeys.

The interior is disappointing, not for its ergonomics, which maintain BMW's traditionally high standards, but because it looks so little different to those of its stable-mates. Such cars need a few design splashes to help create the sense of occasion upon which their existence depends.

Even so, the Z3 2.8 is a fine car. It's not simply the perfect alternative to those who will not wait until the next century to reach the head of the SLK queue, as it has massive appeal of its own. If what you're after is a roadster with traditional values applied in a thoroughly modern but respectful manner, it gets close to the bullseye. The predicted list price is £24,500, going on sale in August.

BMW M ROADSTER

Dial M for

Powered by the magnificent 321bhp engine from the M3, the hot BMW M roadster is capab

Murder

slaughtering most rival sports cars. Peter Robinson guns it

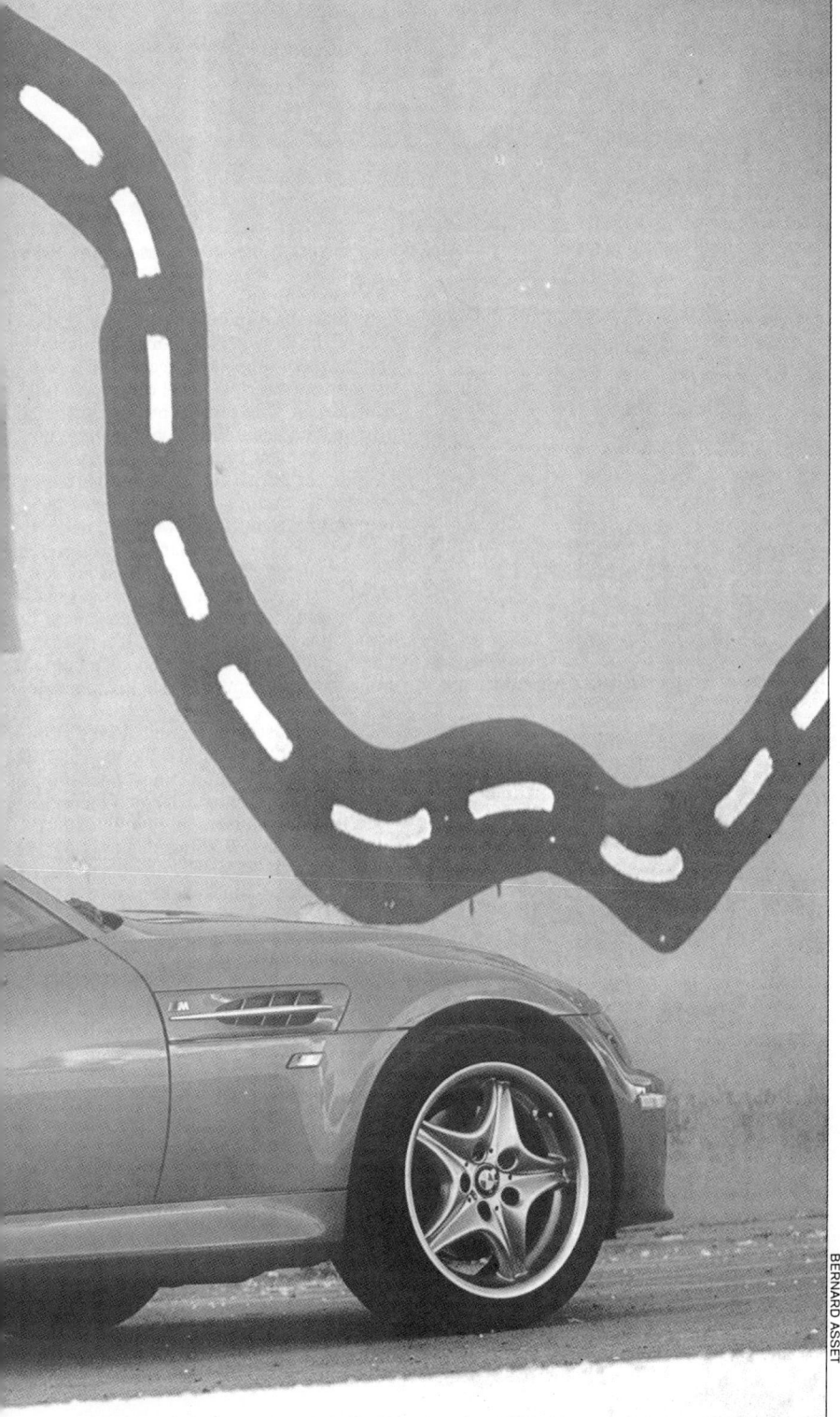

The quickest-accelerating BMW ever is not a Z3, according to BMW. Rather, it's the BMW M roadster. Confused? After all, the obvious name for a marriage of Z3 roadster and M3 power is Z3M; why not that? To understand, you have to drive it, and enjoy thrills that can only be described as seriously sideways, seriously fun.

The M roadster is in a different league from the sports car arena where the Z3 competes. This car aims higher. Much higher. With 321bhp in an engine bay that has always cried out for more, the M roadster has left the world of Porsche Boxsters and Merc SLKs and, with a glimpse over its shoulder at the legendary AC Cobra for inspiration, has gone chasing the Porsche 911 RS and TVR Griffith. The question is, can it catch them?

After a morning on Spanish mountain roads, we finish up at the Jerez circuit. The F1 teams, which come here for winter testing, are absent. It is time to sample that 100bhp-per-litre power and exploit a power-to-weight ratio that guarantees the M roadster's place as the fastest-accelerating production BMW of all time.

On all previous Z3s, even the 2.8, grip always overcomes power, at least in the dry. Not so the M roadster. Bang home the clutch with 5000rpm on the revcounter and be ready to dial in the opposite lock. Through a haze of blue tyre smoke, engine screaming at the 7700rpm cut-out, the BMW explodes away, leaving two long black lines on the tarmac in true Williams FW18 style. And, like contemporary F1 cars, the M roadster is sufficiently old-fashioned to lack electronic traction control.

Few, if any, cars with this near-supercar level of acceleration – BMW claims 5.2sec to 60mph, officially 0.1sec quicker than the M3, and a standing quarter mile in only 13.4sec – are easier to get off the line. For as long as the monster 245/40 ZR17 Michelins hold up, the exercise can effortlessly be duplicated. Power slides? A hoot, once you discover how best to set up the car for a corner and then have the space, and skill, to indulge in the joyous art of balancing power oversteer on the throttle.

We'll admit to doubts about the M roadster when we first heard BMW was planning it. The M3's awesome powerplant under the same bonnet that accepts 115bhp in the basic, admittedly gutless, Z3 1.8 (the one that's fortunately not sold in the UK)? Surely a 179 per cent increase in power would ask too much of any soft-top?

Our apprehension centred around the rear suspension. Like all Z3s, the M roadster inherits the old 3-series semi-trailing arms (also still used by the Compact) and not the more modern and effective multi-links. Such is the thoroughness of BMW development that our misgivings were entirely misplaced.

This may be BMW's hot-rod in terms of outright performance, but that implies the M roadster is crude, even raw. In fact, it behaves impeccably, demanding no more than sensibly judicious use of the loud pedal. It is undoubtedly tighter of structure, sharper of responses and tauter of handling than any of its immediate relatives. If the 1.9 represents the Z3 in simple roadster mode and the 2.8 as a cruiser, the M roadster is pure sports car.

It will take a knowing eye to pick the M roadster, which shares the same muscular haunches as the Z3 2.8. You notice first the front spoiler with its near-horizontal tabs, there to increase aerodynamic downforce. The huge floating caliper, aluminium front discs need plenty of cooling air, so there are no foglights, as on the 2.8, to clutter the outer openings. Satin-finished, cast alloy 17in wheels sit 10mm closer to the body.

Below a new design of rear bumper (which forces the numberplate on to the boot lid and means the BMW propeller is now recessed into the lid's top) sit two pairs of stainless steel exhaust pipes. Immediately behind them are silencers so large that they ensure this is the first BMW not fitted with a spare wheel. There's isn't room. Instead, the boot houses a mini-compressor and quick-forming sealant which should get you to the nearest service centre.

Front suspension, brakes and steering come from the M3. The rear suspension shares the same concept as other Z3s, but every part has been strengthened to cope with the extra torque. According to BMW, the rear subframe is twice as stiff as that of other Z3s. There's even an oil cooler for the rear diff. But alas you can't have an M roadster with the M3's six-speed gearbox, and neither is there the possibility of a sequential gearshift or automatic

Retro-style dash suits M roadster's hairy-chested driving character; new wheel has full-size airbag

The business: but (at last) more power than grip

The best sort of Z3: precise, taut and responsive, M roadster is undiluted sports car. Ride good too

transmission. Again, there's not the room.

A smartened-up interior goes beyond ever that of the 2.8 with brilliantly comfortable and far more supportive seats, and a new three-spoke steering wheel carrying a full-sized airbag. The console brings additional instruments, including an old-fashioned circular ambient temperature gauge. Despite the garish two-tone leather trim, the interior quality is a huge improvement over early Z3s, although we could do without the excessively large, vision-robbing interior mirror. A power-operated hood, not fitted to the pre-production car we drove, will be standard. Belatedly, an optional hard-top is promised for all Z3s for next winter, while twin rollover hoops mounted immediately behind each bucket seat are new options.

The M roadster may weigh 165kg more than the Z3 1.9, but its 1350kg is still 110kg less than the M3 coupe. No wonder the performance is electric. Despite lousy aerodynamics – a best Cd of 0.41 with the hood erect – there's sufficient power for the top speed to be electronically restrained to 155mph, 20mph quicker than the 2.8.

With BMW's double Vanos system adjusting both inlet and exhaust timing, the dohc 3.2-litre in-line six delivers an astonishing band of seamlessly smooth power, from idle to the soft 7700rpm cut-out, 300rpm beyond the red line. Real shove-in-the-back power starts from as low as 2300rpm. Only an uncomfortable resonance when accelerating in fourth from 2500rpm to 2800rpm spoils the refinement.

From 5500rpm, the revcounter needle sprints around the dial and there's always power enough to mask the widely spaced gear ratios. The engine sounds great from outside the car; inside it, with the roof down, wind roar drowns all.

Based upon the official combined fuel figure of 25.4mpg, you might expect the puny 51-litre fuel tank to offer sufficient

Anyone for oversteer? Yes, there's lots, but it's not vicious. And M roadster's responsiveness makes balancing it on the throttle the stuff of driving dreams

range, just. The potential is there, certainly, but people who buy the M roadster are not interested in driving for economy. A range of 150 miles is closer to reality.

Dynamically, the M roadster is brilliant. It needs respect, sure, but for its amazing ride quality, obvious body stiffness, handling adjustability and welcome consistency of responses, this is easily the best Z3. Simply, it points.

While there is never the cohesive balance of a Boxster's chassis to bond car and driver, this BMW is precise, turning into corners with gentle understeer until the driver decides to offset the pushing nose by applying power. Trying harder means being less smooth, but it's great fun. You can sense the inside rear wheel losing traction on tight corners. Back off, though, and the tail hardly moves off line.

Oversteer is not of the vicious lift-off variety but power dependent. The brakes need heat to reach their brilliant best, but that's hardly a weakness.

We drove the M roadster in the same week that BMW GB finally launched right-hand-drive versions of the 1.9. Happily, the M roadster won't be as long in coming. It's expected to go on sale early next year at around £40,000, or £5000 more than a Griffith, but close to four grand below the M3 cabriolet. Now the RS 911 is dead, look no further.

Mighty M: 321bhp meets Z3 for electrifying result

Old but okay: rear semi-trailing arms beefed up

Front suspension, steering and brakes are all taken from the M3 saloon

FACTFILE

BMW M ROADSTER

HOW MUCH?
Estimated price £40,000 On sale January 1998

HOW FAST?
0-60mph 5.2sec Top speed 155mph

HOW THIRSTY?
Urban 17.0mpg Extra urban 35.7mpg Combined 25.4mpg

HOW BIG?
Length 4025mm Width 1740mm Height 1266mm
Wheelbase 2459mm Weight 1350kg
Fuel tank 51 litres (11.2 galls)

ENGINE
Layout 6 cyls in line, 3201cc Max power 321bhp at 7400rpm
Max torque 258lb ft at 3250rpm Specific output 100bhp per litre Power to weight ratio 238bhp per tonne
Installation longitudinal, front, rear-wheel drive
Made of alloy head, cast iron block Bore/stroke 86.4/91mm
Compression ratio 11.3:1 Valve gear 4 per cyl, dohc
Ignition and fuel Electronic digital engine management system

GEARBOX
Type 5-speed manual
Ratios/mph per 1000rpm
1st 4.2/5.5 2nd 2.49/9.4 3rd 1.66/14.0 4th 1.24/18.8
5th 1.00/23.2 Final drive ratio 3.15

SUSPENSION
Front struts, coil springs, anti-roll bar
Rear semi-trailing arms, coil springs, anti-roll bar

STEERING
Type rack and pinion, power assisted
Lock to lock 3.2 turns

BRAKES
Front 315mm ventilated discs Rear 312mm ventilated discs
Anti-lock standard

WHEELS AND TYRES
Size 7.5Jx17 front, 9Jx17 rear Made of cast aluminium alloy
Tyres 225/45 ZR17 front, 245/40 ZR17 rear

All manufacturer's figures

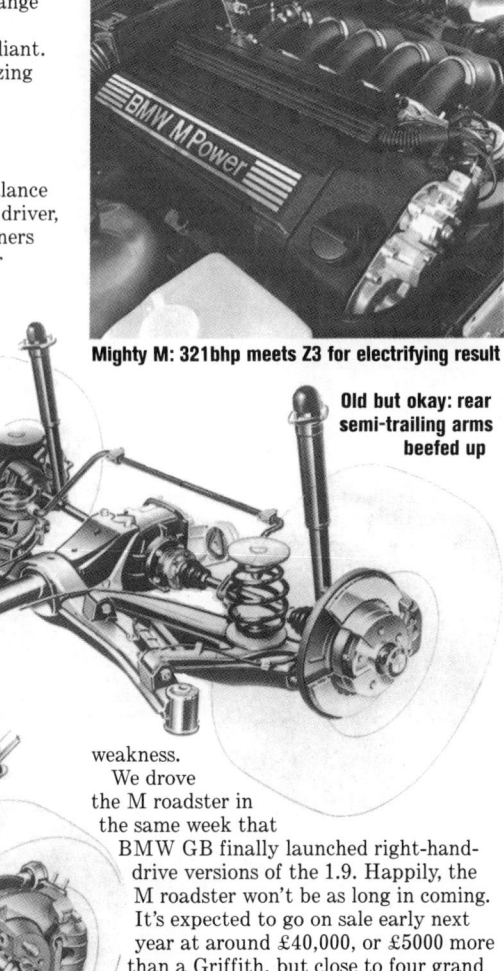

A POTENT 189hp SIX TURNS THIS

Power. It's what separates the winners from the wannabes. And while BMW's luscious Z3 is an indisputable sales and image-building success, those among us who like to burn rubber, thwart Porsches, and hit 90 mph by the top of the freeway onramp were left having to make a few excuses for the underpowered James Bond–mobile.

That all changes as of today.

With their new concoction called the Z3 2.8, BMW engineers have taken the Carroll Shelby approach to roadster design: Stuff in a high-performance powerplant, plaster on some beefy rubber, and let the action begin.

The motive force comes from the same 2.8-liter DOHC inline-six used in BMW's 3 and 5 Series, pumping out 189 horses at 5300 rpm and a bawdy 203 pound-feet of torque at 3950 rpm (down slightly from the other cars' output due to exhaust-system differences). That lets you feel the thrill of 51 more horsepower and 70 additional pound-feet of torque as compared with the 1.9-liter four-cylinder-equipped Z3—a move that increases the car's fun-factor exponentially.

Run 'em off side by side, and the 2.8 mauls its little brother like a starving alligator, blasting 0-60 mph in 6.3 seconds (versus 7.9) and extending the charge all the way to a top speed of 135 mph, while the four-banger runs out of pep at about 120 mph. The power increase will be immediately felt by anyone with more driving brio than an expired jar of Velveeta, as the six's broad torque band moves you rapidly away from a stop, up craggy hills, and into fast-moving traffic without the tach needle pointing any higher than about 3000 rpm. So, while the base Z3 owner must gnash his/her teeth all the way to redline to keep ahead of the madding crowd, all you have to do is dip into the 2.8's deep pool of delicious torque (80 percent of which is available from 1500 rpm).

Because the all-aluminum six extends far rearward in the engine compartment and weighs just 348 pounds, the Z3's front/rear weight distribution actually improves from 52/48 percent to the optimum 50/50 balance. Combined with a 2.6-inch-wider rear track, optional 17-inch wheels and tires, and standard ASC+T (Automatic Stability Control and Traction) with a 25 percent locking rear differential, the Z3's handling acumen is taken to a higher, more controllable level.

Glued to the road by its available 225/45ZR17 (front) and 245/40ZR17

BMW Z3
ROAD TEST
2.8

PRETTY-BOY INTO A PERFORMER

by C. Van Tune
PHOTOGRAPHY BY THE AUTHOR

Compact yet comfy, the Z3's cockpit handles six-footers without complaint. Leather upholstery is standard on the 2.8 model, but if you want the optional chrome trim it will set you back $150.

The Z3's hood opens menacingly to reveal its potent new 2.8-liter/189-horse DOHC six. Choose the five-speed manual and you'll rage to 60 mph in 6.3 seconds; 6.7 seconds with the four-speed automatic.

(rear) Michelin Pilot HX radials, the car we drove with glee up, down, and across the volcanic Atlantic island of Madeira left us with new-found religion. It combines all the inherent goodness of a standard Z3 with a bonus pack of extra abilities you didn't know were missing. With the extra power comes the opportunity to take better advantage of the suspension's grip in uphill charges, dice and squirt your way through traffic, and flash that sadistic grin after blowing off some smart-ass who tried to cheat past you in the gutter lane before the line of parked cars begins. And that's just part of the fun. You'll revel in the superb chassis balance, the heroic brakes, and world-conquering feel of the power rack-and-pinion steering. All of this proves the Z3 2.8 is not just one of the world's most efficient playtoys, it's a damn fine personal transportation module as well.

BMW's designers resisted the temptation to festoon their pumped and cut new model with the boastful badgework of a World Wrestling Federation champion. Not even a demure chrome "2.8" emblem stands on the rear deck to subtly tell the tale. But look more closely and you'll see different fender arches framing the widened rear stance, a slightly more aggressive front fascia, enlarged kidney grilles, and dual exhaust tips. Inside, the well-finished two-seater differs from its less-opulent sibling by delivering standard leather upholstery and power windows, and offering chrome trim around the gauges, door handles, and such. As an extra benefit, an available flip-up windblocker screen resides behind the seats and does a yeoman's job of quelling untoward tonsorial muss with the top down, even at highway speeds.

Like every one of the 35,000 Z3s already delivered to worldwide buyers, the 2.8 version is built in Spartanburg, South Carolina. And at a base price of $35,900 (just $6475 more than the base four-cylinder version), the hottest BMW roadster is a standout value, ready to duke it out with Porsche's Boxster for sportster supremacy. Aah, but there's even better news for the future: We can almost taste the lusty thrill that 1998's 240-horsepower M Roadster will deliver.

That Bond sissy won't know what hit him. **MT**

TECH DATA
BMW Z3 2.8

GENERAL/POWERTRAIN
Importer	BMW of North America, Inc., Woodcliff Lake, N.J.
Location of final assembly plant	Spartanburg, S.C.
Body style	2-door, 2-passenger
EPA size class	Two-seat convertible
Drivetrain layout	Front engine, rear drive
Airbag	Dual
Engine configuration	I-6, DOHC, 4 valves/cylinder
Engine displacement, ci/cc	170/2793
Horsepower, hp @ rpm, SAE net	189 @ 5300
Torque, lb-ft @ rpm, SAE net	203 @ 3950
Transmission	5-speed manual

DIMENSIONS
Wheelbase, in./mm	96.3/2446
Track, f/r, in./mm	55.6/58.8/1413/1494
Length, in./mm	158.5/4026
Width, in./mm	68.5/1740
Height, in./mm	50.9/1293
Ground clearance, in./mm	4.6/117
Mfr's base curb weight, lb	2778
Weight distribution, f/r, %	50/50
Cargo capacity, cu ft	6.2
Fuel capacity, gal.	13.5
Weight/power ratio, lb/hp	14.7

CHASSIS
Suspension, f/r	MacPherson struts, coil springs, anti-roll bar/semi-trailing arms, coil springs, anti-roll bar
Steering	Rack and pinion, power assist
Turning circle, ft	39.4
Brakes, f/r	Vented discs/discs, ABS
Wheels, f/r, in.	17 x 7.5/17 x 8.5, aluminum alloy
Tires, f/r	Michelin Pilot HX, 225/45ZR17/245/40ZR17

PERFORMANCE
Acceleration, 0–60 mph, sec	6.3
Standing quarter mile, sec/mph	14.8/91.4
Braking, 60–0, ft	115.1
Handling, lateral acceleration, g	0.87
Speed through 600-ft slalom, mph	66.9
EPA fuel economy, mpg, city/hwy	19/27

PRICE
Base price	$35,900
Price as tested	$37,620

http://www.bmwusa.com

BMW Z3

Practice your James Bond driving style and bring plenty of hair gel, because the BMW Z3 commands attention like no other machine in its price range. Our wish for more power will be answered soon by the M3's inline-six.

Two-seat sports cars come in all flavors and price ranges, and this one's a steal at its $28,750 base sticker. But, don't make the mistake of only comparing the BMW Z3 to a Mazda Miata—match it up with the head-turning appeal of a $135,000 Ferrari F355 Spider, as well. It only takes one drive through Beverly Hills, Malibu, or along Hollywood Boulevard to be made keenly aware of this ragtop's phenomenal gawk factor.

On the scale where a Bruce and Demi sighting only rates a gawk factor of "2," and bags of money falling from the sky command a mere "7," a single pass by a bus stop in the Z3 pegs the needle with a solid "10." Even without its prerelease publicity as the latest James Bond getaway machine, this Bimmer could hardly be more conspicuous if it were filled with circus mimes, dipped in gelignite, and launched from atop Bob Stupak's Stratosphere hotel in Las Vegas.

OK, so it isn't a good choice for the bashful; there are plenty of old Novas and Tercels for those people to drive. What you want is something to match your outgoing personality—something with the searing good looks of the BMW Z3.

Although only about three inches longer stem-to-stern than a Miata, the Z3 rides atop a 7.1-inch-longer wheelbase. This added stretch gives the BMW less of a Matchbox car sort of feel and melds with the masterful frontal styling to create an ambiance of a much larger, more expensive sportster.

And bystanders aren't at all shy about leaning in and asking questions. So be prepared. Here are the answers you'll need to know: 1.) Yes, this is the James Bond car; 2.) No, but I'm Pierce's stunt double; 3.) It's a 1.9-liter/138-horse DOHC four; 4.) Zero to 60 in 7.9 seconds; 5.) Only about $30,000; 6.) Yes, 30 thousand, not 70; 7.) Sure, you can have a ride.

Answers to the questions they're not likely to ask, but you should know anyway are: 1.) It's built in Spartanburg, South Carolina; 2.) It matches the skidpad grip of a Ferrari 355 Spider, at a brain-sloshing 0.93 g; 3.) The manual top can be lowered without you having to leave the driver's seat; 4.) There's 70 percent greater trunk space than in a Miata (but even that only gives you 6.2 cubic feet); 5.) It's 2.3 mph faster through *MT*'s slalom course (at 69.5 mph) than a new Porsche 911 Targa; 6.) The cockpit is comfy for two six-footers; 7.) Unfortunately, what the looks portend, the acceleration can't match.

For most Z3 gawkers, that last item won't matter, because this is a car they truly believe is the neatest thing on four wheels. And for around $30,000, nothing can match it. —*C. Van Tune*

Long-Term BMW Z3

Panache in the extreme, and power mostly in your dreams.
But, hey, it's still a Bimmer.

BY STEVE SPENCE

"Hysteria" probably overstates the reaction, but parking a BMW Z3 roadster curbside one Saturday night back in March of 1996 along Miami's South Beach—the car had yet to go on sale—created an instantaneous envy and saliva festival.

It's 15 months later, and the Z3 still gets gawks, but now you don't have to pry yuppies off the hood with a spatula. That is now the predicament of Porsche Boxster owners and perhaps a few Mercedes-Benz SLK owners with rural addresses.

A major attraction of the Z3 is its $29,320 base price, and that includes power for the windows, door locks, side mirrors, and seats, plus air conditioning and cruise control. Our car totaled $31,827 after we opted for beige-leather upholstery ($1150), heated seats ($500), a metallic Montreal Blue paint job ($475), the on-board computer that reveals fuel economy, range, outside temperature, and so on ($300), and two floor mats at $41 apiece.

We were not thrilled by the promise of the powerplant, a 1.8-liter four from the 3-series bin that had been bored and stroked to 1.9 liters. In full-shout mode, it put out an unadventurous-sounding 138 horses and 133 pound-feet of torque. (More like it is the 2.8 six-cylinder that appeared within a year.)

BMW has never offered the sort of thunderous muffler music produced by some American cars to announce their macho, but that didn't stop the crew here

52

Here's proof that Canada has beaches—this one's called Grand Bend, in Ontario. At right, a view of the Z3 interior before it got the aftermarket wheel and shifter knob.

Rants and Raves

The car's nose is a work of art. I was surprised and delighted at the arc that stretches from the wheelhouse, around the nose, and back to the wheelhouse. Simply gorgeous—not a vertical or horizontal surface anywhere.
—Schroeder

What a hot car! On the way to Canada, we got more stares than the Pope in a Speedo.
—Mosher

Lack of torque is annoying. Really have to punch it just to spin the tires, even on wet surfaces. And how about a tilting steering wheel?
—Davidek

This Z3 shares one trait with the 3-series. Hinge friction (doors and trunk) is extreme, making slamming necessary to ensure closure.
—Bedard

I drove the Z3 for 2170 miles, and with the top up, it doesn't have the feel of a sports car—much more comfort than one would expect. My co-pilot was six foot six, and the car had enough room for him.
—Vaughn

Tach and temp gauge finally back in operation—this really is a traditional sports car, complete with imitation Lucas electrics.
—Bedard

from complaining about our Z3's lack of a "distinct exhaust note." Phillips pointed out the "amusing Lilliputian exhaust note. Nearly silent, unlike a Miata; then, at high revs, it sounds like, well, a Tercel. Weird." Idzikowski spoke for many: "Way too wimpy."

You couldn't stop the Miata comparisons, the enthusiast stereotypes here immediately siding with that Japanese throwback to English sports-car simplicity. A comparably equipped MX-5 Miata is $24,215, although the Z3 is roomier, quieter, more refined, and, folks, it's a Bimmer.

The radio wasn't up to Bimmer standards. Its speakers are tinny. Schroeder put it this way: "Radio blows. Blow it outta there. Someone call Alpine."

Some found the manual top easy to use, others complained when it took more than one arm to raise or lower it. The rear window, made of plastic and unzippable, seemed murky to look through from the start, and complaints would grow.

But soon it was summer, and our Z3 was busy, indeed. This car tells you when it wants its oil changed. On the instrument panel just below the odometer readout, five green LED lights in a row drop off one by one as the miles go by; when they're all gone, a yellow one appears, along with one of two announcements: "Oil Service" or "Inspection." Ignore it, and a red light appears. Our first call for fresh oil came on July 10, at 9552 miles. A Bloomfield Hills dealer added six quarts of oil and a filter for just $24, although we were hit with an hour's labor ($68) and a can of Krex oil treatment ($5) that we hadn't asked for. One oil change: $97. (Do-it-yourselfers will need a $50 tool to

53

reset the computer's maintenance clock.) Maintenance alternates between a simple oil change and a lengthier full inspection at intervals that, for us, occurred roughly every 9500 miles.

That summer, the air conditioner could barely keep up. After four days in the shop, a new condenser was ordered, which was installed a month later at 18,328 miles during the car's first "Inspection I" service. That included a huge list of look-sees and cost $202, plus $15 for six quarts of fresh oil, $2.50 for washer solvent, and $9 for two wiper blades (these are tallied under normal wear). The bill was $247.

We began having a peculiar handling problem. "Hit a bump midway through a curve, and the Z3's all over the road," said Schroeder. Said another tester: "Cornering is fantastic, but if you encounter bumps in the middle of hard cornering or braking, hang on!" BMW had heard the same complaints on some early cars. It found that the rear ride height of our car, which was supposed to be 22.76 inches (measured from the top of the wheel-well arch to the bottom of the wheel's rim, with a full tank and 330 pounds' worth of passengers), was in fact 22.11 inches. Two springs were replaced without charge in early November.

As another dreadful Michigan winter closed in, the Z3's popularity waned. Two pea-sized fray spots appeared on the convertible top at the edge of one supporting rod. The door locks were sticky, fussy to open. There were rattles in the cockpit. On February 2, the fuel gauge went cuckoo; the tank would be full, but the gauge might show a quarter-tank, and when the driver tried to fill up, he'd find it already full. No fun in 18-degree weather. The dealer replaced the sending unit and fuel

Z3 goes antique hunting in Ohio (left). Meanwhile, babes Jan and Jackie enliven prissy Ann Arbor scene.

pump under warranty. By February 18, the driver's door lock wouldn't budge, so the dealer replaced both door handles and lock assemblies and worked on the rattles. We were back to the dealer on March 3 (at 25,676 miles) with the fuel gauge still lying to us. The dealer yanked the entire instrument cluster and replaced it. All fixes were under warranty.

Meanwhile, we improved the Z3's appeal on ice with four Yokohama AVS S4 snow tires ($636), which boosted grip substantially. Up here in the tundra, these tires are a necessity.

Shipped off to Bedard's hideout in Florida in March (27,500 miles), the "Check Engine" warning light began flashing on and off for no reason, then the tachometer freaked out, bouncing all over the place. This weirdness coincided with the yellow dash light ordering an oil service, but a dealer in Palm Harbor stiffed us by performing an "Inspection II" service, which was $203. It needed the oil and filter swap, but it was way early for new plugs, tranny oil, an air cleaner, and other parts. We ate a $381 bill. (To keep the accounting straight, we ordered a normal oil change at the next service interval.)

The Florida dealer decided the car's schizophrenic warning lights were caused by a failed oxygen sensor. By the time the ordered parts came in we were back in Michigan, where the Z3 spent four days away from home in April while the dealer replaced the hazard switch and the oxygen sensor. The only way to cure the crazy tach was to—gulp!—replace the entire instrument cluster again. All work was covered by the warranty.

On Cinco de Mayo, at 38,954 miles, our Z3 got a final $83 oil change. Some cad here had also crunched the front air dam and bumper cover, and replacing them cost us $865. Service costs over almost 40,000 miles cost $799, about four times what our $17,168 long-term Miata cost in 1991 over 30,000 miles. Our blue Z3 got 27 miles to the gallon. The Boxster and the SLK will be pricey to service, too, although neither is a throwback to rough-hewn sports cars the way the Z3 is.

Our test car was quicker a year later. Where it had once turned 0 to 60 in 8.0 seconds and 0 to 100 in 26.4 seconds, it now did them in 7.8 and 24.5, and the quarter-mile time dropped from 16.2 to 16.0. The Z3 also stopped nine feet earlier from 70 to 0 mph (169 feet new, 160 feet at 40,000 miles).

The Z3 was a popular ride here. It's sleek, an attention getter, a civilizing treatment of the roughhouse sports-car ideal. We prefer the 2.8-liter six model, but it sends the price heavenward by more than $7000, to $36,508. Still, that's about five grand cheaper than the Boxster and the SLK.

Prom night in Allen Park, Michigan: Meghan and Karen ditch their dates for our Z3.

BMW Z3
Vehicle type: front-engine, rear-wheel-drive, 2-passenger, 2-door convertible

Price as tested: $31,827 (base price: $29,320)

Engine type: DOHC 16-valve 4-in-line, iron block and aluminum head, Bosch HFM Motronic M5.2 engine-control system with port fuel injection

Displacement . . . . 116 cu in, 1895cc
Power (SAE net) . . . . 138 bhp @ 6000 rpm
Torque (SAE net) . . . . 133 lb-ft @ 4300 rpm
Transmission . . . . 5-speed manual
Wheelbase . . . . 96.3 in
Length . . . . 158.5 in
Curb weight . . . . 2760 lb

Performance:	new	40,000
Zero to 60 mph	8.0 sec	7.8 sec
Zero to 100 mph	26.4 sec	24.5 sec
Street start, 5–60 mph	9.6 sec	9.0 sec
Standing 1/4-mile	16.2 sec	16.0 sec
	@ 83 mph	@ 84 mph
Braking, 70–0 mph	169 ft	160 ft
Roadholding, 300-ft-dia skidpad	0.86 g	0.87 g
Top speed (governor limited)	116 mph	116 mph

EPA fuel economy, city driving . . . . 23 mpg
C/D observed fuel economy . . . . 27 mpg
Unscheduled oil additions . . . . 0 qt

Service and repair stops:
Scheduled 4
Unscheduled 6

Operating costs (for 40,000 miles):
Service $799
Normal wear $9
Repair $865
Gasoline (@ $1.17 per gallon) $1738

Life expectancies (extrapolated from 40,000-mile test):
Tires . . . . 42,000 miles
Front brake pads . . . . 100,000 miles
Rear brake pads . . . . more than 100,000 miles

Baubles and Bolt-Ons

Maybe you'd better hear this sitting down.

The tony steering wheel you see here, sold by BMW and installed on our test car, cost $650. The matching wood shifter is $98. Smelling salts, anyone? About half the wheel, which is fatter in diameter than the stock version and finger-grooved underneath, is burled walnut, the other half covered by grippy leather at the "nine" and "three" hand positions.

Worth it? Head techy Markus says yes, swearing he'd "happily skip the power seat and seat heaters to pay for them." But Jon Davis, who drove the Z3 some 5500 miles through the One Lap of America course, said the spokes are set in the wrong positions and dismissed the steering wheel as "ugly."

The rowdies around here complained fervently about the "wimpy" exhaust note. We did our gosh darnedest to get B & B Fabrication of Glendale, Arizona, to send us one of its Z3 muffler and tailpipe units but discovered they're available only for the 2.8-liter model. (For info, call 888–228–7435. The system is less than the wheel—about $600, we're told.)

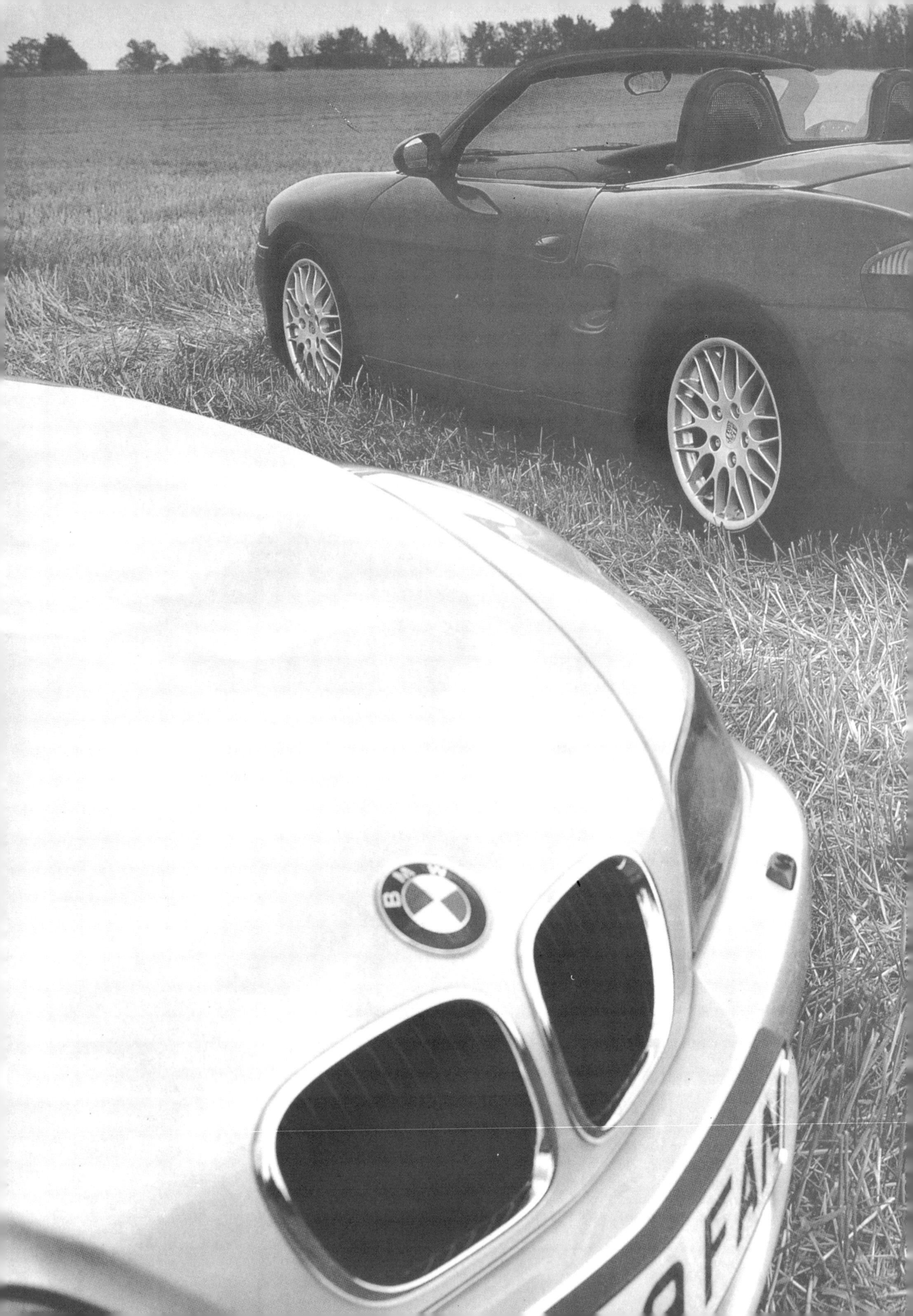

AUTOCAR TWIN TEST NO 4275

BMW Z3 vs Porsche Boxster

MODELS TESTED Z3 2.8/Boxster
PRICES £28,115/£34,095 **TOP SPEED** 134/139mph
0-60mph 6.7/6.5sec **30-70mph** 5.9/6.6sec
60-0mph 2.6/2.6sec **OVERALL MPG** 24.9/26.8

T he justification for this showdown, we admit, seems rather slim. All the pre-fight form tells us that Porsche's brilliant Boxster – a roadster we've dubbed the greatest on the planet – will administer a serious beating to BMW's simpler, cruder Z3, even newly minted with the company's classic 2.8-litre straight six under the lid.

But BMW has upset the apple cart before. Inspiration missing from lesser models has a knack of cropping up when more cylinders and power climb on board. The 3-series Compact, for instance, has been utterly transformed by a baby six. And, from our time behind the wheel so far, we know that the six-cylinder Z3 is essentially a sports car reborn, packing the punch and panache that the 1.9-litre four-cylinder version so badly lacks.

So the question is this. Is the BMW Z3's great engine, keen pricing and enticing bundle of standard kit enough to rattle the purely hewn, mid-engined Porsche Boxster?

Here are its best shots. One, it's nearly £6000 cheaper. Two, it totes leather-covered, electrically operated seats and traction control – all extras on the Porsche. Three, it has more torque. And not everyone is mad about mid-engined cars. A strong traditionalist streak persists in the sports car sector and it's hard to think of a car that serves it better than the Z3 2.8.

Nose to tail

With a sweet 2.8-litre six under its long, traditional bonnet, the BMW Z3 now has the firepower to go head to head with the bespoke mid-engined Boxster

TOM SALT

DESIGN & ENGINEERING

The contrasts are stark. Porsche's aim with the Boxster wasn't to pander to the market conventions in many ways exemplified by the Z3, but to follow its own inimitable star. So the 911 lineage is allowed to shine through, even though it's the Boxster that forms the basis for Porsche's future. Lovers of the old 911 won't be alienated by the Boxster body's aversion to anything that resembles a straight line. Some have called its styling blob-like and bloated, but it's amazing the difference a colour can make. Our test car's red paintwork revealed hitherto unsuspected curves. It looked great, and it turned heads.

Most obvious departures are the placement of the engine – from behind to in front of the rear axle – and the use of water instead of air to cool it. Perhaps more remarkable is the up-sizing, not just from the exquisite 1993 Detroit show concept car, but also from the 911 itself. The Boxster is both longer and wider than its progenitor, though most of the extra length manifests itself as overhang and, of course, with the engine slung longitudinally, there's room for only two seats. But the 996 (new 911) is bigger still and at least those meaty overhangs help packaging; the Boxster's fore and aft luggage compartments aren't just surprisingly big, they're also usefully shaped.

Packaging considerations also ditched the textbook wishbones in favour of struts, but the whiff of compromise wasn't allowed to drift anywhere near the all-new water-cooled 2.5-litre flat six engine with its four camshafts, 24 valves, VarioCam valve timing and dry sump. A peak of 204bhp is fair considering the engine's modest swept volume, but 181lb ft of torque is nothing to write home about – even if 81 per cent of the maximum is available from just 1750rpm.

If the Boxster is vulnerable to attack anywhere from the Z3, it's here. BMW's stalwart 2.8 straight six may not have the exotic configuration or high-tech gloss of Porsche's new showcase powerplant, but it's got more beef than a Bovril factory. True, 192bhp is 12bhp shy of the Boxster's peak output, but 202lb ft of torque is arguably the more important figure when it comes to driveability, and the in-gear flexibility figures bear it out.

Proportions and layout of Z3's cabin have a traditional feel; driving position is comfortable, but too high

BMW Z3 2.8 1 Instruments are neat, attractive, but a bit dull beside the Boxster's 2 Five-speed gearbox has pleasingly crisp action, but again isn't as crisp as the Porsche's 3 Dashboard is carried over virtually unchanged from the Compact, although touches of chrome around instruments succeed in lifting it visually 4 Passenger airbag is a £480 extra 5 Wheel is height adjustable 6 Driving position is higher, more saloon like than the Boxster's

RSCHE BOXSTER 1 RS-badged instruments are perhaps the best touch about Boxster's handsome oard 2 Centre console, incredibly, costs an extra £210. Stowage space is limited to small doorbins without it tion control is a pricy £850 option, whereas the Z3's is a standard fitment 4 Leather seats of the test car add a ning £2387 to the price 5 Steering wheel is fixed in just the right position 6 Air con costs a further £1850

r's cabin shows evidence of strong ideas having been watered down. Comfy, roomy and sporty

Much of the Z3's grass roots engineering comes from the humble 3-series Compact, itself a development of the previous generation Three. That includes the semi-trailing arm rear suspension responsible for educating an entire generation of drivers in the finer art of opposite lock.

A criticism of the comparatively tame 1.9 Z3 is that its tail is, if anything, too firmly glued to the black stuff. Everything else being equal, the 2.8's grunt would probably sort that. But everything else isn't equal. The rear track has been broadened by 16mm and the 16in alloys wear gumball 225/50 section tyres. The semi-trailing arms have been stiffened, too, and there are larger anti-roll bars front and rear.

Purist Porsche outguns more basic BMW
BMW ★★★
Porsche ★★★★★

PERFORMANCE, BRAKES

Figures are one thing, feel quite another. Against the clock, it's a nip 'n' tuck contest. The Boxster holds an edge in outright power and weight, the Z3 hits harder through the mid-range and has slightly shorter overall gearing. All in all, it's a pretty close call.

The one area where the Porsche does put some daylight between itself and its adversary is flat out. It lapped the Millbrook bowl at an average of 139mph. The Z3 rushes up to 130mph then slowly battles on to a maximum of 134mph.

From a standing start, things are a lot tighter. With traction controls switched out, both cars launch well, but it's the Boxster, with 54 per cent of its weight over the rear wheels, that punches to 30mph harder, posting an exceptional 2.2sec against the Z3's 2.4sec.

It's an advantage the Porsche never relinquishes. Sixty up and it's still 0.2sec ahead on 6.5sec, and from there on it gradually pulls away. By 100mph the BMW is struggling to stay in touch; 18.4sec is swift by any standards, but the Boxster is there in 18sec dead.

If extracting every last ounce of energy from an engine is your thing, the Porsche is your car. The 2.5 boxer engine spins so sweetly and eagerly to its 6600rpm limit that it makes even the BMW's silky straight six sound slightly raw, uncouth. Imagine a 911 soundtrack digitally remastered for extra

SPECIFICATIONS

BMW Z3 2.8

ENGINE
- **Layout** 6 cyls in line, 2793cc
- **Max power** 192bhp at 5300rpm
- **Max torque** 202lb ft at 3950rpm
- **Specific output** 68bhp per litre
- **Power to weight** 143bhp per tonne
- **Torque to weight** 151lb ft per tonne
- **Installation** Front, longitudinal, rear-wheel drive
- **Construction** Aluminium alloy head and block
- **Bore/stroke** 84/84mm
- **Valve gear** 4 valves per cyl, dohc
- **Compression ratio** 10.2:1
- **Ignition and fuel** Siemens MS41.0 engine management

GEARBOX
- **Type** 5-speed manual
- **Ratios/mph per 1000rpm**
- **1st** 4.20/5.4 **2nd** 2.49/9.1
- **3rd** 1.66/13.6 **4th** 1.24/18.2
- **5th** 1.00/22.7
- **Final drive ratio** 3.15

SUSPENSION
- **Front** MacPherson struts, coil springs, anti-roll bar
- **Rear** Semi-trailing arms, coil springs, anti-roll bar

STEERING
- **Type** Rack and pinion, power assisted
- **Turns lock to lock** 2.9 turns

BRAKES
- **Front** 286mm ventilated discs
- **Rear** 272mm discs
- **Anti-lock** Standard

WHEELS AND TYRES
- **Wheel size** 7Jx16in
- **Made of** Light alloy
- **Tyres** 225/50 R16 Michelin Pilot MXXV SX
- **Spare** Space saver

VITAL STATISTICS
Body 2dr roadster **Cd** 0.42 **Front/rear tracks** 1413/1494mm **Turning circle** 10.0m **Min/max front leg room** 980/1160mm **Max front head room** 940mm **Interior width** 1310mm **Min/max boot width** 1000/1170mm **Boot height** 490mm **Boot length** 780mm **Loading height** 640mm **VDA boot volume** 165 litres/dm³ **Kerb weight** 1335kg (claimed) **Distribution f/r** n/a **Max payload** 790kg **Max towing weight** n/a

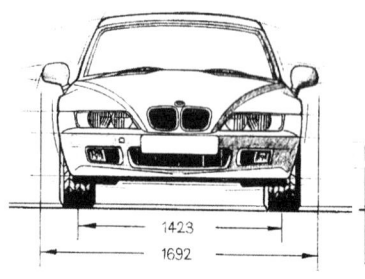

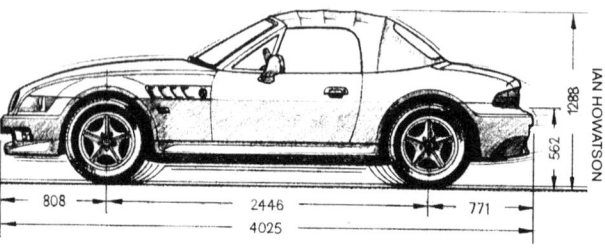

PORSCHE BOXSTER

ENGINE
- **Layout** 6 cyl boxer, 2480cc
- **Max power** 204bhp at 6000rpm
- **Max torque** 181lb ft at 4500rpm
- **Specific output** 82bhp per litre
- **Power to weight** 159bhp per tonne
- **Torque to weight** 141lb ft per tonne
- **Installation** mid, longitudinal, rear-wheel drive
- **Construction** aluminium alloy head and block
- **Bore/stroke** 86/72mm
- **Valve gear** 4 valves per cyl, dohc
- **Compression ratio** 11.0:1
- **Ignition and fuel** Bosch Motronic 5.2

GEARBOX
- **Type** 5-speed manual
- **Ratios/mph per 1000rpm**
- **1st** 3.50/5.4 **2nd** 2.12/8.9
- **3rd** 1.43/13.3 **4th** 1.03/18.4
- **5th** 0.79/24.0 **Final drive ratio** 3.88

SUSPENSION
- **Front** Struts, coils, longitudinal and transverse arms, anti-roll bar
- **Rear** Struts, coils, longitudinal and transverse arms, anti-roll bar

STEERING
- **Type** Rack & pinion, power assisted
- **Turns lock to lock** 3.0

BRAKES
- **Front** 298mm ventilated discs
- **Rear** 292mm ventilated discs
- **Anti-lock** standard

WHEELS AND TYRES
- **Wheel size** 6Jx16in (f), 7Jx16in (r) (optional 17in wheels fitted)
- **Made of** cast alloy
- **Tyres** 205/55 ZR16 (f) 225/50 ZR16 (r) (optional 205/50 ZR17 front and 255/40 ZR17s fitted)
- **Spare** Space saver

VITAL STATISTICS
Body 2dr roadster **Cd** 0.31 **Front/rear tracks** 1455/1508mm **Turning circle** 10.9m **Min/max leg room** 863/1092mm **Min/max head room** 889/927mm **Interior width** 1346mm **Min/max boot width** 737/1143mm **Boot height front/rear** 584/330mm **Load height front/rear** 584/737mm **VDA boot volume front/rear combined** 254 litres/dm³ **Kerb weight** 1242kg **Distribution f/r** 46/54 per cent **Max payload** n/a **Max towing weight** n/a

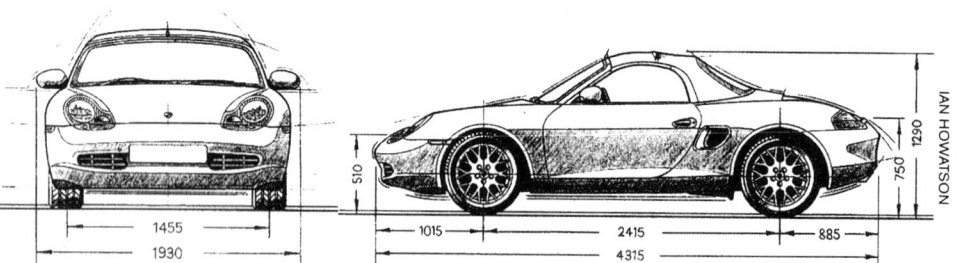

FUEL CONSUMPTION
TEST RESULTS (mpg: Average, Touring, Best, Worst)
- BMW Z3
- Porsche Boxster

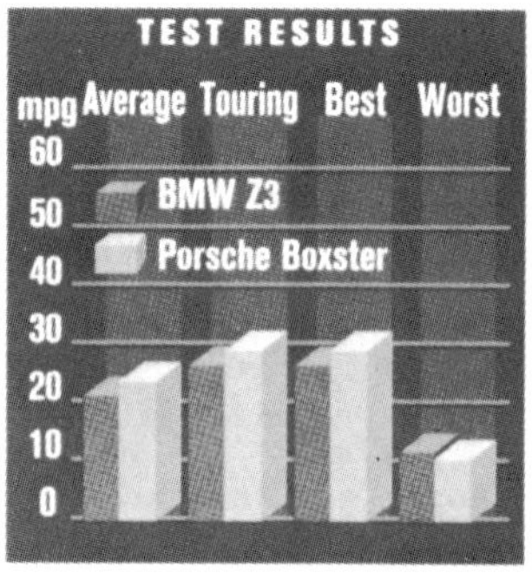

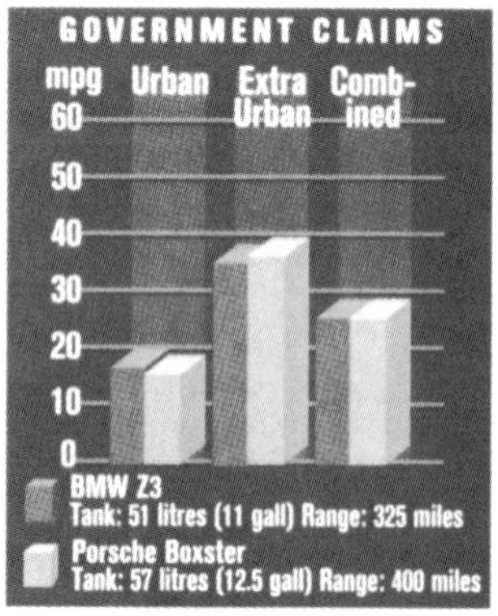

GOVERNMENT CLAIMS
(mpg: Urban, Extra Urban, Combined)
- BMW Z3 — Tank: 51 litres (11 gall) Range: 325 miles
- Porsche Boxster — Tank: 57 litres (12.5 gall) Range: 400 miles

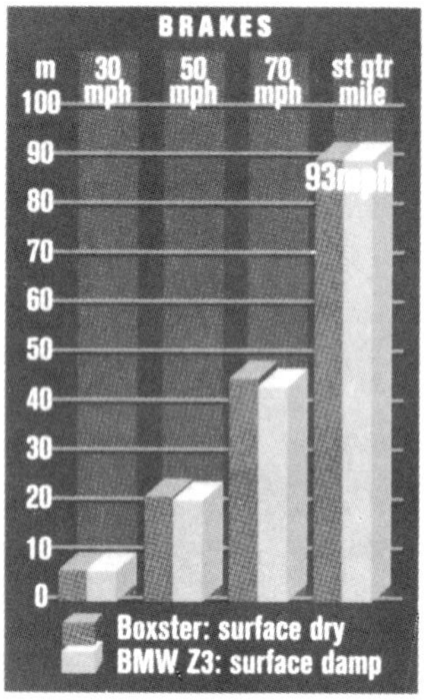

BRAKES
(m, 30mph, 50mph, 70mph, st qtr mile — 93mph)
- Boxster: surface dry
- BMW Z3: surface damp

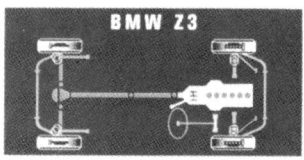

The performance figures were taken at the **Millbrook Proving Ground** with the BMW Z3's odometer reading 3265 miles and the Porsche Boxster's odometer reading 2177 miles. AUTOCAR test results are protected by world copyright and may not be reproduced without the editor's written permission.

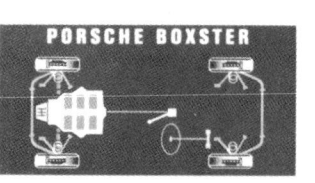

AUTOCAR road tests are conducted using BP Unleaded or BP Diesel Plus with additives to help keep engines cleaner

clarity and smoothness and you more or less have it. But neglect to change down or call it at day at 5800rpm and the Z3 will have you. You won't, though. The Boxster's accurate and light short-throw shift is a joy, the engine at its most musical on the boil.

Taking the Z3 to the red line has its benefits, too – and its snappy gearchange is almost as rewarding – but it isn't the prerequisite for rapid progress it is in the Porsche. Even in fifth it can muster plenty of overtaking urge, zipping between 50 and 70mph in just 8.3sec. Try the same thing in the Boxster (10.7sec) and Z3 drivers will start drumming their fingers. It's the same low down: between 30 and 50mph in fourth the Z3 outhauls the Boxster to the tune of 1.1sec with a time of 6.1sec. It simply feels lustier. But not as special.

Few cars get the better of a Porsche when it comes to stopping, and the Z3 2.8 isn't one of them. Its anchors are firm, fade-free and more than up to the performance. The Boxster's are simply fantastic.

Both are terrific – in different ways
BMW ★★★★
Porsche ★★★★

ECONOMY

Muscular engines without much weight to lug around are usually a surefire formula for surprising fuel economy. Bring a slippery shape and sensibly long-legged overall gearing to the party (the Boxster turns up with a Cd of just 0.31 and pulls 24mph/1000rpm in top) and the picture starts to look quite rosy.

During its time with us, the Porsche usually managed to return around 25mpg, no matter how hard it was driven, and it comfortably drifted into the low 30s when we eased back. This gives it a practical range of just over 400 miles, even from its smallish 57-litre tank.

The Z3 couldn't quite match this, managing only 24.9mpg overall. It wasn't quite as thrifty at touring either, struggling to break the 30mpg barrier. Unfortunately, its 51-litre tank is even smaller than the Boxster's and limits its range to around 325 miles.

Both do well, but Boxster has the edge
BMW ★★★
Porsche ★★★★

Z3's boot is distinctly weekend-sized. Door bin cubby useful for coins. Big grip, but oversteer on demand

Deeper of Porsche's two boots is up front. Door sill levers open both. On limit, Boxster will oversteer, just

HANDLING & RIDE

The Boxster appears to hold all the cards. If you're going to build a truly modern sports car – which the Porsche undoubtedly is – you don't even think about putting the engine anywhere but in the middle. And, having decided you're going to do that, using a "flat" engine, with a low centre of gravity, is better still. Squat and broad-shouldered as the Z3 is, it's comparatively narrow and tall next to the Boxster. Just by looking at them – and even the Porsche's flat belly has been designed to reduce lift – you can tell the car from Stuttgart will be faster through any combination of corners.

Take it as read. But it's the cars' different interpretations of what adds up to a good time that fascinates. The Z3 is largely a what-you-see-is-what-you-get merchant. Big six and gearbox in the nose driving to the rear wheels, fat tyres, effective but not intrusive traction control – the same basic ingredients as the big Healeys and ACs of the '50s and '60s.

The arrangement feels familiar, simple and honest, and you know what to expect the moment you point its long nose into the first bend: understeer. How much depends on your entry speed and what you do next with the steering and the throttle. Kill the gas, help along the natural tuck-in with a little more lock and jump back on the power (traction control off, of course) and you can neutralise the understeer and maybe kick the tail out a few degrees. Exaggerate the sequence and you might even coax the Z3 into a full-blooded power slide. Good, indulgent fun if you're on the right road in the right mood.

It's when you're on the wrong road (bumpy) and in a hurry that the Z3 is less convincing. Its steering isn't especially good – plenty of feel but not enough straight-ahead precision – and its body starts to shudder and shimmy, as if the grippy tyres are generating forces the body structure can't quite handle. It's here that the Z3 starts to feel rough round the edges.

And it's here that the Boxster is at its sublime best. In any situation it feels an altogether sharper tool than the BMW, marrying even more grip with crisper turn-in, swifter responses, purer and more informative steering and a flatter ride. Moreover, its bodyshell feels absolutely rigid – no different, in fact, to a hard-topped Porsche's.

But there is an argument that the Boxster is almost too good. Simple techniques that can get the Z3 sliding are ignored. A conciliatory nudge of oversteer is as much as even the most determined pilot can hope for. Roll on 3.2 power.

Z3 can be fun, but Boxster is inspired
BMW ★★★
Porsche ★★★★★

HOW THEY COMPARE

	BMW Z3 2.8	**PORSCHE BOXSTER**
PERFORMANCE		
Maximum speeds		
5th	134mph/5900rpm	139mph/5800rpm
4th	114/6250	122/6600
3rd	85/6250	89/6600
2nd	57/6250	59/6600
1st	34/6250	35/6600
Acceleration through the gears		
0-30mph	2.4sec	2.2sec
0-40	3.6	3.5
0-50	5.0	4.9
0-60	6.7	6.5
0-70	8.3	8.8
0-80	11.3	11.1
0-90	14.4	14.1
0-100	18.4	18.0
0-110	24.1	22.4
0-120	32.2	28.6
30-70	5.9	6.6
Standing quarter	15.2sec/93mph	15.6sec/90mph
Standing kilometre	27.5sec/114mph	28.5sec/115mph
Acceleration in third/fourth/fifth		
10-30mph	5.1/–/–	5.7/–/–
20-40	4.6/6.7/8.8	4.9/7.8/12.2
30-50	4.2/6.1/8.3	4.7/7.2/10.5
40-60	4.0/6.1/8.0	4.5/7.0/10.4
50-70	4.2/6.0/8.3	4.3/7.2/10.7
60-80	4.5/5.9/8.5	4.4/7.4/11.3
70-90	–/6.5/8.8	5.6/7.4/12.3
80-100	–/7.4/9.9	–/7.6/13.4
90-110	–/9.3/11.3	–/8.3/14.0
100-120	–/–/–	–/10.6/15.9
Braking: 60-0mph	2.6sec	2.6sec
COSTS		
On-road price	£28,115	£34,095
Price as tested	£29,180	£41,203
Interim service	8000 miles	12,000 miles
Major service	16,000 miles	24,000 miles
Insurance group	16	18
Warranty	3 years/60,000 miles	2 years/unlimited miles
	6 years anti-corrosion	10 years anti-corrosion
EQUIPMENT		
Automatic transmission	£1040	£2600
Airbag driver/passenger	●/**£480**	●/●
Metallic paint	£360	£745
Rollover hoops	£360	●
Centre console	●	£210
Anti-lock brakes	●	●
17in alloy wheels	£790	£1075
Alarm/immobiliser	●/●	●/●
RDS stereo/CD player	●/**£150**	£485
Traction control	●	£850
Leather seats	●	£2387
Electric hood	●	●
Air conditioning	£1100	£1850
Heated seats	£295	£347
Trip computer	●	£280
Headlight washers	£225	£195
Wind deflector	£250	£251
● standard – not available **bold type** denotes option fitted to test car		
SOLD BY		
	BMW GB Ltd, Bracknell, Berks Tel: 01344 426565	Porsche Cars GB Ltd, Reading, Berks Tel: 01734 303666

Z3 2.8 has nothing to fear from Boxster when it comes to overtaking

COMFORT & EQUIPMENT

Neither car stints on basic comfort or kit. Both have comfortable, well-shaped seats with a wide span of adjustments, well thought-out driving positions and conveniently placed controls. But only the Z3 gets power seat adjustment as standard. Likewise traction control and leather seats.

You feel as if you're sitting lower in the Boxster, as if its cabin is broader and roomier – which, indeed, it is. And its windscreen is shallow and set a long way forward; you don't feel the header rail is about to bash your bonce.

The BMW's deeper screen is good in this respect, too, but the cabin ambience is entirely different – narrower and more traditional. Its pimply leather seats don't hug your body quite as securely, but the relationship between the seating position and the scuttle is pleasingly low-slung and snug.

For design, neither wows the driver in the way the Mercedes SLK does. But if the Boxster disappoints, it's more through comparison with the exquisitely detailed Detroit concept car than the Z3. At least the residue of strong ideas leaves the Boxster's cabin with the whiff of originality, and the tightly clustered instrument dials with their "RS" graphics are a genuinely nice touch.

The Z3's facia is merely generic BMW – neat, ergonomically sound, clearly instrumented – and wouldn't seem out of place in a small saloon or hatch. The chromed controls and gear knob add some sparkle (literally) but don't make it any more sporty.

One of the reasons the Porsche costs more is that its powered hood is faster, more slickly engineered and less hassle – almost as much of a talking point, as witnessed by roadside observers, as the SLK's. It generates less wind noise at speed when erect than the Z3's, too, though the amount of buffeting with it down is similarly low.

Extra kit tips scales in Z3's favour
BMW ★★★★
Porsche ★★★

Z3's control starts to waver on bumpy roads; not as rigid as Boxster

-valve straight six engine has a better spread of torque and more of it

MARKET & FINANCE

Just weeks after the model went on sale, used BMW Z3 2.8s outnumber used Porsche Boxsters by two to one in the nation's classifieds. Dial in the more prevalent Z3 1.9 – which, apart from its slightly narrower rear wheel arches, looks no different from the 2.8 – and the ratio of used Z3s to used Boxsters becomes five to one. That's a lot of Z3s, but only a few Boxsters.

The fact is, the Boxster will always be a more rare and coveted sight than the 10-a-penny Z3. That's important to punters who shop at these rarified heights and means that the Boxster will be the more desirable and shrewd purchase.

But we're splitting hairs. The Z3 still packs the badge, the looks and the quality to hold a tight second place behind the Boxster, now and in the future. Against the Boxster, it is also much better value for money. Leather, essential in any prestige car these days, is standard in the Z3 2.8 but a whopping £2387 option in the Boxster.

Those other essentials, air conditioning and metallic paint, cost £1100 and £360 respectively on the Z3, £1850 and £745 on the Boxster.

In standard trim, the Z3 costs £6000 less than the Boxster. Specify the BMW with air con and metallic paint, the Porsche with leather, air con and metallic paint, and that £6000 price gap grows to £10,000, the Z3 costing £29,575, the Boxster £39,077.

For clean-living types, insurance is a relatively painless experience. We called the new prestige and performance car arm of Bennetts insurance, called Bennetts Elite (0990 202090), for quotes. For a 36-year-old office manager living in Surrey, with a clean licence and full no claims, BE quoted a premium of £484 with a £275 excess for the Boxster, and £338 and a £200 excess for the Z3.

Boxster is rarer, but Z3 is better value
BMW ★★★★
Porsche ★★★★★

Boxster unflappable in the extreme, even with traction control off

THE AUTOCAR VERDICT

OUR CHOICE

Boxster is a more focused and complete car than Z3 – a quality act

HOW THE RIVALS COMPARE

MAKE/MODEL	LIST PRICE	MPH/0-60	TEST DATE
BMW 328i convertible	£31,355	139/6.6sec	26.7.95
Same glorious engine as Z3, but capable of seating four			★★★
Mercedes-Benz SLK	£30,090	140/7.4sec	20.11.96
Superbly designed and built, but not as good to drive			★★★★
TVR Chimaera 4.0	£30,650	158/5.2sec	16.6.93
The hairiest open top car on sale at the price, it's a blast			★★★★

The Boxster's reputation is barely bruised by this scrap. As we wait for the new 911 to arrive in November, it's the most complete and capable car Porsche makes and far too good for the Z3 – even one with a smooth and brawny 2.8-litre straight six under the bonnet and a £6000 price advantage acting in its favour. But more on that in a moment.

Considering what the Z3 is – essentially a 3-series Compact with a roadster body and a lot of fine tuning – it does a decent job. Top speed apart, it's very nearly as quick as the Porsche from a standing start and serves up more in-gear punch for overtaking. If its cabin isn't the last word in sporting chic, it is practical, comfortable, quite roomy and well laid out. And when you look at the long list of standard equipment (leather, power seats, traction control, trip computer and so on), its value isn't in question. In fact, it's here more than anywhere else that it hurts the Boxster. Porsche even makes you pay for the centre console.

But the Boxster is an easy car to forgive. Even standing still it has the BMW licked, with the styling and finish to make it look even more expensive than the price difference would lead you to believe, a hood that deserves a place in a gallery of kinetic art, and two boots rather than one.

On the move, the gap grows alarmingly. Not only is the Boxster's body structure much stiffer than the Z3's, but its dynamic resolve is stronger, too. It turns with more acuity, hangs on longer, rides bumps and ruts with more control and less shudder, and brakes as only a Porsche really can. It is both more focused and more complete. Admittedly, in the wet, or blasting out of a tight second-gear bend, the Z3 can be a little more fun, its rear tyres giving up the fight more readily than the Porsche's. The Boxster will hang its tail out, too, but you'll be travelling faster when it does and will need to be both swifter and more accurate with corrective lock.

In the end, it isn't so much what the Boxster does but the quality way it does it. This is a great Porsche.

TEST NOTES

On dry, smoothly surfaced roads, you'd be hard pushed to tell (apart from the warning light in the instrument panel) when the Boxster's traction control is switched off. Provoking wheelspin when cornering is next to impossible – a compelling demonstration of the mid-engined Porsche's tremendous traction.

The BMW's power hood has a double latching system. Once undone, you have to push the leading edge of the roof away from the windscreen before you hit the button and step outside to fit the tonneau when it's finished folding. The Boxster has an ingenious one-handed latch system and a significantly faster fold time. And it disappears under a solid cover.

BMW Z3 2.8 ★★★ **Porsche Boxster** ★★★★★
Porsche shows BMW how to build a complete sports car

63

ROAD TEST No 4300

BMW Z3

MODEL TESTED M roadster **ON-ROAD PRICE** £40,570
TOP SPEED 155mph **0-60MPH** 5.1sec
30-70MPH 4.4sec **60-0MPH** 2.8sec **MPG** 20.4
FOR Sublime engine, muscle car looks, quickest BMW yet
AGAINST Dull handling, links with basic Z3, driving position

The concept behind the new BMW M roadster is as simple as it is seductive. Take the pretty but ineffectual Z3 roadster, cram the 321bhp engine from the M3 under the bonnet and hang on tight.

A fine idea in anybody's book but not quite as simple as it sounds. Remember that by BMW standards the Z3 is not a very sophisticated machine. It's built on the bones of the Compact, which itself borrows much of its hardware from the previous generation 3-series.

No matter. It's exactly the fillip that the Z3 needs. Forget the 1.9-litre car with its mild-mannered engine. The M roadster is the genuine article, a hairy-chested hot rod complete with over-inflated wheel arches, engine and, of course, price. At £40,570 it is nearly twice the price of the basic Z3.

Dials, mirrors exclusive to M car

To justify such an enormous leap in cost, BMW has done everything in its power to distance the M roadster from its lesser brethren. The swollen flanks may be shared with the 2.8-litre Z3, but the gaping front air dam, repositioned number plate, wing mirrors, chromed side strakes, clear indicator lenses and quad exhausts are all new. As if that wasn't enough, the colour-keyed roof and most of the paint options are also exclusive.

It's the same story inside the car, with contrasting coloured leather panels on the seats, doors and dashboard. Three new circular dials also adorn the centre console, along with chrome highlights to the gear gate and ventilation controls. It's a good effort and far more comprehensive than the transformation from 328i to M3, but it still doesn't feel as special as a purpose-built machine like a TVR.

More problematic for some will be the limited size of the cockpit. The standard powered seats slide back just far enough to accommodate the legs of a 6ft driver, but the fixed steering wheel doesn't allow for any fine tuning of the driving position. The low waistline of the doors also leaves you feeling rather exposed, even with the seats on their lowest setting.

The electrically powered hood slides up and down with typical BMW efficiency, but there is no extra lining to improve its interior looks or refinement and you still have to fit the flexible

Oversteer on demand in wet, but understeer more common on dry roads

STAN PAPIOR

Tonneau tucks away in slim boot

64

plastic tonneau cover by hand. Not a problem in a £20,000 car but a little out of place in a £40,000 one.

Still, the M roadster is meant to be about fun rather than luxury, and in performance terms the M roadster more than fulfils its brief.

The 3.2-litre straight six from the M3 has already been acknowledged as one of the greats. Not only does it develop over 100bhp per litre (thanks to its double Vanos variable valve timing) and rev to 7700rpm, but it is also extraordinarily smooth, flexible and economical. We averaged 20.4mpg, but nearer 30mpg should be possible on a long run if you eke out the contents of

Compressor replaces spare tyre

BMW has tried hard to make M roadster feel more special than other Z3s

65

PERFORMANCE AND SPECIFICATIONS

ENGINE
Layout 6 cyls in line, 3201cc
Max power 321bhp at 7400rpm
Max torque 258lb ft at 3250rpm
Specific output 100bhp per litre
Power to weight 229bhp per tonne
Torque to weight 184lb ft per tonne
Installation Front, longitudinal, rear drive
Construction Aluminium head, iron block
Bore/stroke 86.4/91mm
Valve gear 4 per cyl, dohc
Compression ratio 11.3:1
Ignition and fuel MS S50 Digital Motor Electronics management

GEARBOX
Type 5-speed manual
Ratios/mph per 1000rpm
1st 4.21/5.5 **2nd** 2.49/9.3 **3rd** 1.66/14.0
4th 1.24/18.8 **5th** 1.00/23.2
Final drive ratio 3.15:1

MAXIMUM SPEEDS
5th 155mph/6700rpm **4th** 141/7500
3rd 105/7500 **2nd** 70/7500 **1st** 41/7500

ACCELERATION FROM REST
True mph	sec	speedo mph
30	2.1	31
40	2.9	41
50	4.0	52
60	5.1	62
70	6.5	73
80	8.2	83
90	9.9	94
100	12.0	104
110	15.0	114
120	18.2	125
130	22.4	135

Standing qtr mile 13.6sec/106mph
Standing km 24.3sec/134mph
30-70mph through gears 4.4sec

ACCELERATION IN GEARS
mph	5th	4th	3rd	2nd
10-30	–	–	4.3	2.8
20-40	7.5	5.7	3.6	2.2
30-50	6.9	5.0	3.3	2.2
40-60	6.2	4.8	3.3	2.2
50-70	6.2	4.6	3.3	–
60-80	6.0	4.6	3.4	–
70-90	6.2	4.8	3.5	–
80-100	6.7	5.5	3.8	–
90-110	7.3	5.6	–	–
100-120	8.3	6.2	–	–

STEERING
Type Rack and pinion, power assisted **Turns lock to lock** 3.2

CONTROLS IN DETAIL
1 M roadster uses five-speed gearbox rather than M3's six-speeder
2 Heated seats are standard 3 New analogue dials for clock, engine oil temperature and exterior temperature 4 Contrasting leather inserts unique to M roadster 5 Special M wheel is smaller than normal but no adjustment is possible 6 Powered seats are attractive and supportive

SUSPENSION
Front Struts, lower wishbones, coil springs, dampers, anti-roll bar
Rear Semi-trailing arms, coil springs, dampers, anti-roll bar

WHEELS & TYRES
Wheel size 7.5Jx17in (f), 9Jx17 (r)
Made of Cast alloy **Tyres** 225/45 ZR17(f), 245/40 ZR17(r) Michelin Pilot SX
Spare Compressor and sealant

BRAKES
Front 315mm ventilated discs
Rear 312mm ventilated discs
Anti-lock Standard

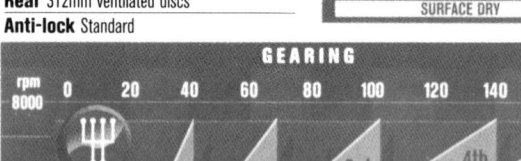

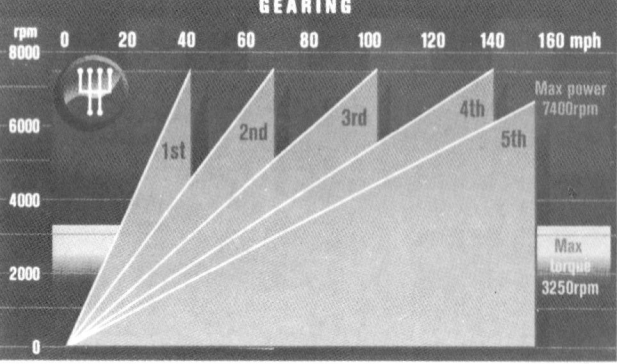

BRAKES (test)
60-0mph: 2.8sec
30mph: 9.4m; 50mph: 24.8m; 70mph: 49.0m; st qtr mile: 108.3m (106 mph)
SURFACE DRY

GEARING
Max power 7400rpm; Max torque 3250rpm

AUTOCAR road tests are conducted using BP Unleaded or BP Diesel Plus with additives to help keep engines cleaner

FUEL CONSUMPTION
TEST RESULTS (mpg)
Average 20.4 | Touring 26.6 | Best 26.6 | Worst 14.8

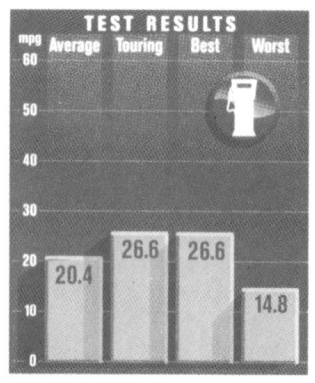

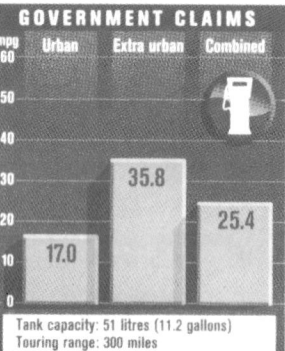

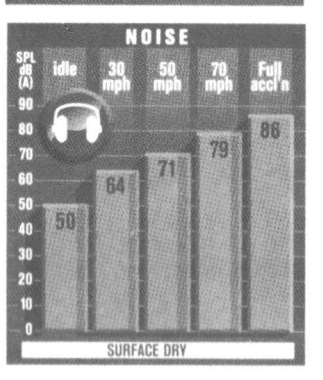

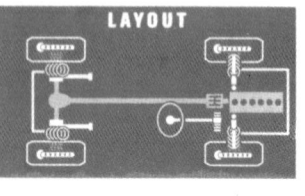

GOVERNMENT CLAIMS (mpg)
Urban 17.0 | Extra urban 35.8 | Combined 25.4

Tank capacity: 51 litres (11.2 gallons)
Touring range: 300 miles

NOISE (SPL dB(A))
idle 50 | 30mph 64 | 50mph 71 | 70mph 79 | Full accl'n 86
SURFACE DRY

LAYOUT

Body 2dr roadster **Cd** 0.42 **Front/rear tracks** 1422/1492mm **Turning circle** 10.4m **Min/max front leg room** 980/1060mm **Max front head room** 940mm **Interior width** 1310mm **Boot length/width/height** 780/1170/490mm **VDA boot volume** 165 litres/dm³ **Kerb weight** 1402kg **Weight distribution front/rear** n/a **Max payload** 790kg **Max towing weight** n/a

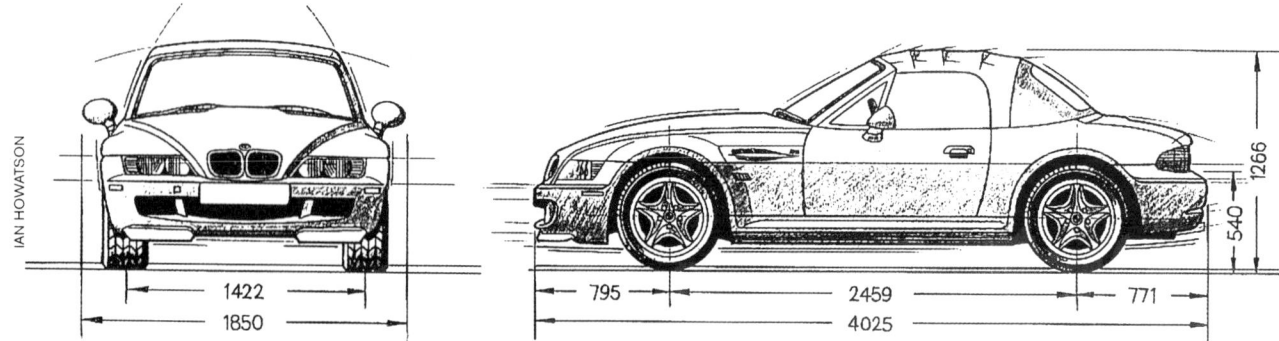

Dimensions: 1422 / 1850 (front); 795 / 2459 / 771, 4025 (side); 540 / 1266 (height)

IAN HOWATSON

The performance figures were taken with the odometer reading 3650 miles. **AUTOCAR** test results are protected by world copyright and may not be reproduced without the editor's written permission

WHAT IT COSTS

On-road price	£40,570
Total as tested	£42,040
Cost per mile	83.3p

EQUIPMENT
(**bold** = options fitted to test car)

Automatic transmission	–
Air conditioning	**£1100**
Heated sports front seats	●
Full leather interior	●
Electric hood	●
Anti-lock brakes	●
Airbag driver/passenger	●/●
Alarm/immobiliser	●/●
Remote central locking	●
RDS stereo/CD player in lieu	●/£175
Rollover bars	●
17in Roadstar alloy wheels	●
Metallic paint	**£370**

● standard – not available

Insurance group 20

WARRANTY
3 years unconditional

SERVICING
Minor 7500 miles, 0.3hrs, parts £10.96
Major 15,000 miles, 3hrs, parts £38.45

M roadster's handling is more rewarding than that of lesser Z3s, but it lacks feel and finesse of true sports car

M roadster exhilarating in straight line – the most accelerative BMW yet

◆ its miserly 51-litre tank. The engine's low-down torque (258lb ft at 3250rpm) negates the jump in weight from 1164kg in the 1.9-litre Z3 to 1402kg in the M roadster, but it's hardly in the sports car spirit.

Light up the tyres from rest, snatch second gear and 5.1sec later the M roadster hits 60mph. Keep the pedal flattened, grab third and in another 6.9sec you're past 100mph. The standing quarter and kilometre are dispensed with in equally nonchalant fashion after 13.6sec and 24.3sec respectively, and so it goes on until the M roadster bumps into its electronic speed limiter at exactly 155mph, making this the most accelerative production car BMW has ever built.

Given this performance, you might expect the M roadster to feel unruly, but in truth it is more pussycat than tomcat. Like the M3, you have to take care co-ordinating throttle and clutch to avoid driveline shunt at low speeds, but this aside it's as easy to drive as a 1.6-litre Compact. The clutch is a little heavier, the gears a little snappier, but the experience is still eerily familiar. Not so the compound disc brakes from the M3, which carve off speed with an authority rarely found on road cars. The anti-lock sensors never cut in a second too soon and even the pedal feels exactly as it should: firm, consistent and very reassuring.

Given BMW's normal pre-occupation with safety, it is surprising to find that this is the only Z3 not fitted with electronic traction control. In its place is a more enthusiast-orientated limited slip differential. It's still possible to unsettle the rear end on a dry road, but it takes commitment. It's another matter on a damp one, when it's all too easy to spin the wheels inadvertently on the exit of a roundabout and kick the tail out of line.

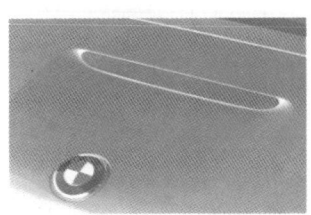

Shorter springs, stiffer dampers, thicker anti-roll bars and reinforced suspension components raise the M roadster's handling onto a different plane from other Z3s, but there is no getting away from the limitations of its comparatively flexible chassis. It's fine on smooth, grippy roads, but the body still shimmies over the worst ruts and bumps. Perhaps next year's Z3-based coupe will provide the answer. The steering, although nicely weighted, lacks the speed, precision and feel of a truly incisive sports car. Turn-in feels heavy-handed next to a Porsche Boxster or Lotus Elise and the huge amount of lateral grip generated by the stiff-walled 245/40 ZR17 Michelins comes at the expense of a more throttle-sensitive handling balance. Unless you set it up perfectly at the entrance to the corner, even heavy throttle applications will just push the M roadster's nose wide.

It will still cover distances at a prodigious rate, and even the most hardened drivers cannot fail to be impressed by its astonishing brakes, performance

Chrome fins hark back to BMW 507

Badge moved up to brake light

and grip. But the type of driving experience that draws you into the car, hypnotises you with its involvement and keeps you glued to the seat until long after dark still eludes the M roadster. Despite all its technical brilliance and the audacity of BMW in building such an outrageous machine, we can't see many people falling in love with this car in the way that it's so easy to do with an Elise, Boxster or TVR Chimaera. This is just a shade disappointing from a car with so much to boast about.

Storming 321bhp engine makes light work of Z3

★★★
Brutally fast but uninvolving sports car

[NEWCOMERS]

BMW M3 Roadster

'At last, a spiritual successor to the great

Barlow uses quick rack and 321bhp to defy laws of physics. Car entered stage right, incidentally

M-powered for thrills

BMW's much-criticised Z3 finally gets the muscle to truly impress

THERE'S *COLD* IN THEM thar hills, and on the hidden edges of Sussex's serpentine back roads, ice is making progress unpredictable. Time to tread warily. The M Roadster has no traction control. The M Roadster has 321bhp. Over to Peter Loecker, one of a trio of enigmatic men BMW (UK) has lured away from the shadowy world of BMW M (GmbH) to talk double VANOS and big numbers through the back wheels. The original power broker, in fact.

'No, not power. Yes, of course we like lots of power. But for profit everyday, it is necessary to maximise not power but torque.'

No wonder he looks so smug: there's no shortage of either facility in this most potent of Z3s, finally on sale in the UK and equipped with enough firepower to mount an effective salvage operation on the damaged reputation of BMW's roadster. Damaged? Not as far as BMW itself is concerned: the 100,000th Z3 to roll off the Spartanburg production line was recently delivered to a Yorkshire butcher, while more than 4500 of them have found grateful homes in the rest of the UK since its arrival a year ago. But the 1.9-litre Z3 is a pretty limp-wristed device, and many of its basic dynamic flaws (slow steering, lack of agility, a general wobbliness) are exacerbated rather than eliminated by the introduction of BMW's otherwise lovely 193bhp 2.8-litre engine. Suddenly, the powerhouse M Roadster we've long been promised begins to sound like a less tantalising proposition.

The men from M, however, do not mess about. Their take on the Z3 is off to a flying start because it looks like a Z3 should: brawny, arches bulging (they're 86mm wider at the rear) with the right size rubber (225/45ZR 17 at the front, 245/40s out back) wrapped around deeply sculpted M alloys, the whole thing in M guise squatting 10mm lower. Some people still don't like its piscine facial expression or long nose/stubby tail profile, but at least it doesn't look like some toy-town sports car.

Inside the well-stocked cabin, there's a very, er, German interpretation of style: the dash and seat inserts are trimmed in body-colour leather, while the instruments sit in chromed bezels. All very teutonically retro. Aberrations of taste aside, the M Roadster's cabin is comfortably intimate, though six-footers will find it a bit cramped.

Only a madman could find anything wrong with the M's engine, though; with the roof stashed (a simple, electrically powered process), the bass-heavy, throaty rumble that's long made the M3 so aurally stimulating is even more seductive in its smaller roadster sibling, a 24-valve wall of sound filtering through four fat tail-pipes. If it goes as well as it sounds... There are a few hurdles, however: the Z3's semi-trailing arm rear suspension is clearly a cruder set-up than the M3's clever multi-link arrangement, though stiffer springs, dampers and roll-bars all round ought to compensate.

So you take it easy to begin with, reacquainting yourself with this engine's astonishing tractability, so lusty that it'll pull in a gently gathering surge from 1800rpm in fifth gear. Herr Loecker can relax. And despite all that extra grunt, the gearchange is swift and precise, the clutch perfectly weighted, the littlest M feeling incredibly (almost disappointingly) well-mannered despite its bulging potential. Then you nail it and it throws off its creamy six-cylinder languor in a snarling display of arrogance,

Ms that have gone before'

Chromed bezels give a retro feel, accommodation cramped for six-footers

Just 400 Ms are coming to the UK this year, so hurry if you want one

smashing past 60mph in just over five seconds on its way to that familiarly governed 155mph max. Jesus. That lightweight benchmark the Elise still feels more urgent up to a point, but there's no substitute for power, and few things could match (or catch) the M, even on an icy B-road. It even rides well.

Nor does it let go easily, in the dry at least, though a vague rear-end shimmy over uneven surfaces is a reminder of the rear suspension's relative crudity. Give it a big bootful in first or second gear – out of a junction, into a corner, wherever – and the M does its thing in dramatic but progressive style; catching the slide, meanwhile, is a much more rewarding, less intimidating task than in less potent Z3s thanks to its faster rack.

BMW is importing 400 M Roadsters this year, and even at a hefty £40,000 it'll have no problem shifting them. More so than the current M3, it has a rawness that casts it as the true spiritual successor to the great Ms that have gone before.
JASON BARLOW

BMW M ROADSTER
PRICE: £40,500
ENGINE: 3201cc 24V straight six, 321bhp, 258lb ft
PERFORMANCE: 155mph (governed), 5.2sec 0-60mph, 25.5mpg (EC ave)
ON SALE IN UK: Now

BMW M ROADSTER
Price £40,000 (est)
Date tested 13.8.97

On paper the BMW M Roadster is the answer to sports car enthusiasts' dreams. And an unequivocal response to those who have criticised both 1.9 and 2.8-litre Z3s for their lack of performance. The M3's 321bhp straight-six should do the trick. Not everybody loves the Z3's looks, but even *Autocar*'s non-Z3 fans agreed that the wider arched and more generously tyred – wrapped around wonderful deeply dished alloys – M Roadster looks more handsome and serious than its lower-powered brothers.

One way to prove that M Roadster is only Z3 not to have traction control

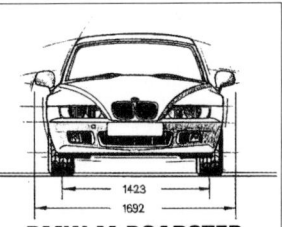

BMW M ROADSTER	
HOW MUCH?	
Price	£40,000 (est)
Test date	13.8.97
HOW FAST?	
0-30mph	2.1sec
0-60mph	5.1sec
0-100mph	12.0sec
30-70mph	4.4sec
Stand qtr mile	13.6sec/106mph
Standing kilo	n/a
30-50mph in 4th	5.0sec
50-70mph in top	6.2sec
Top speed	156mph
60-0mph	2.8sec
Noise at 70mph	n/a
Test weight	1402kg
HOW THIRSTY?	
Overall test average	20.4mpg
Touring route	35.7mpg
ENGINE	
Layout	6-cyls in line, 3201cc
Max power	321bhp at 7400rpm
Max torque	258lb ft at 3250rpm
Specific output	100bhp/litre
Power to weight	229bhp/tonne
GEARBOX	
Type	5-speed manual
Top gear	23.2mph/1000rpm
VERDICT ★★★	
Hugely fast and better looking than its lesser Z3 brethren, but it is still strangely uninvolving for a sports car.	

The car is speed-limited to 156mph, though most would agree that this is adequate for a small roadster. Acceleration is also quite acceptable with a 0-60mph time of 5.1sec.

If raw performance was all that counted in the making of a great sports car, then BMW would have a winner in its stable. But it isn't. There's more to the recipe than just brawn. You sit high in the Z3, rather than low down as if you are part of it. The controls are all easy to operate, smooth and progressive. The steering, like that on the M3, would benefit from more feel and precision.

It is amazing that a car with the M Roadster's power and performance could be uninvolving, but it is. Certainly, for £40,000 the M Roadster provides a reward smaller than the sum of its parts.

321bhp engine limited to 155mph

Seats are too high for a roadster

69

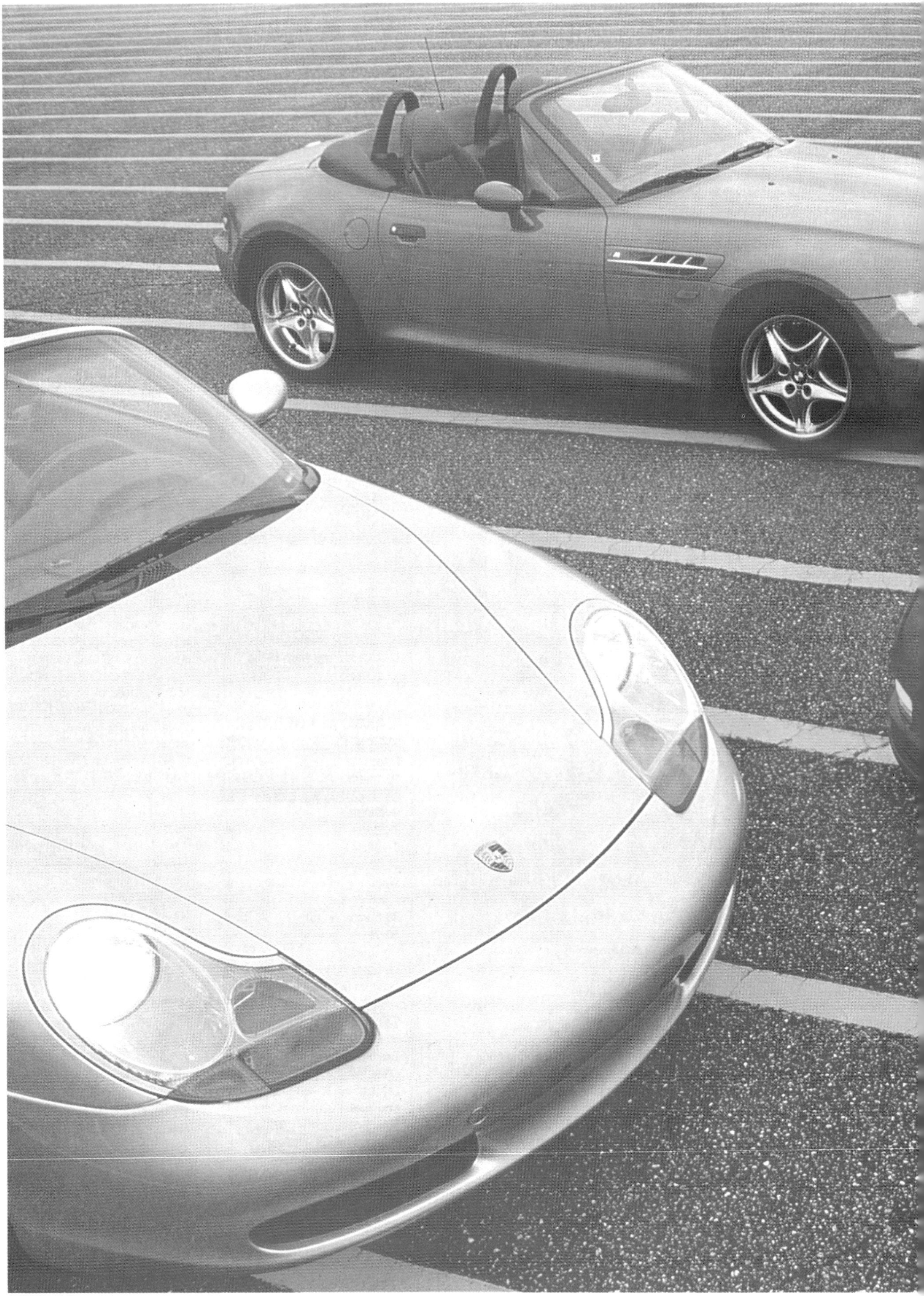

COMPARISON TEST

Road Hunks

Parsimony be damned.
When you're talking power,
performance, and pizzazz,
you're talking big ticket.

BY PHIL BERG

In the past 27 months, five new roadsters have been introduced in the United States. The sportiest of the five are the Porsche Boxster (which went on sale in January 1997), the BMW Z3 (which beat the Boxster to the showroom by a full year), and the Chevrolet Corvette, which has been available since last September. The two other roadsters, which are not included in this comparison test, are the Mercedes SLK230, introduced in February 1996, and Mazda's latest Miata, which should be widely available as you read this. The soft-riding, automatic-only Benz is a luxury cruiser first and is notably slower and less athletic than the three ruffians gathered here, although they challenge it nearly dollar for dollar in the market. The venerable Miata continues to be a pure roadster, but its abilities are limited by its moderately powered motor, which contributes to its affordable $20,000 price, half the price of those in the group we've gathered.

The Corvette, the Boxster, and the new, more-powerful version of BMW's Z3, called the M roadster, are all-out, go-for-broke roadsters, the performance benchmarks. We voted the Corvette and the Boxster two of our 10Best Cars for 1998 (the M roadster was still in the box). On the latest Corvette, many of the details—lack of rattles, the clean gauges, and extra storage space—are so well executed that you'd think each of these roadsters was hand-built. Porsche's Boxster shares the new 911's front fenders and doors and uses a 201-hp, 2.5-liter version of the watercooled flat-six engine in the new 296-hp, 3.4-liter 911 series. Do we like these cars? Does an owl hoot?

BMW will begin selling the M roadster in April. It's powered by the company's strongest six-cylinder engine, also found in the M3 five-seat coupe and sedan. The engine swap required a thoroughly retuned suspension and steering gear, moving the battery from the right side of the trunk to the center to make room for a dual exhaust, filling the fenders with wider tires, and installing four chunky tailpipes that exit in back.

All this extra work takes place on the same assembly line in Spartanburg, South Carolina, that builds the entire world's supply of BMW roadsters. BMW's famed M (for "motorsport") department was responsible for the design and engineering and sends complete 240-hp motors from Munich, Germany, to the upstate Carolinians for installation. The lower-powered Z3 models' engines come from Austria.

The M roadster's in-line six-cylinder is a rev-happy motor. It fits into the small chassis the way the Great Hot Rodder in the Sky

PHOTOGRAPHY BY DICK KELLEY

Chevrolet Corvette

Highs: Abundant power, balanced handling, and a trunk built for three.

Lows: You must get out of the car to lower and raise the roof, and the car's bulk makes it tough to maneuver in tiny personal spaces.

The Verdict: The fleetest of the pack, but it's larger and feels less connected to the road than the others.

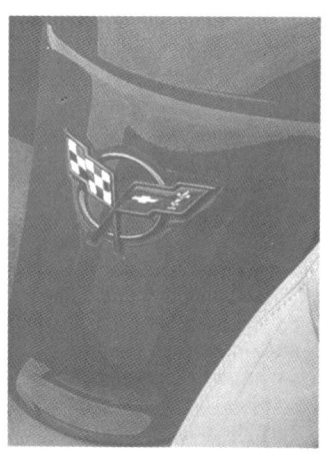

intended, increasing output from the Z3 2.8-liter's 189 hp to 240 hp.

These three roadsters cost big money—$49,235 for the Corvette, $46,385 for the Boxster, and $43,245 for the M roadster as tested. Chevrolet predicts sales of upwards of 10,000 Corvette convertibles in 1998. Porsche hopes to sell 8000 Boxsters, and BMW figures 3500 of the 20,000 Z3s it makes for 1998 will be the M-roadster model.

Exposure to the environment in our three test cars is as undiluted as in any convertible, but roadsters make no compromises for passengers behind the front seats. This selfish seating arrangement focuses the abilities of the Corvette, the Boxster, and the M roadster on pleasing the driver. Which one is the best at it?

3rd Place
Chevrolet Corvette

We'd drive the Corvette roadster about 40 percent of the time with its top down. Chalk that up to its interior spaciousness—there's a lot of room inside for wind to swirl around and give you a chill. Although the stereo volume automatically rises with increased speed, the driveline and the exhaust make enough noise that you often need to withdraw from driving on challenging roads and just aim the car down a straight road, letting your senses cool down.

The good news is the noises are good noises: Throaty, hot-rod-style burbling and provocative "back pops" (just quieter than a full-fledged backfire) from the exhaust make playing with the throttle fun for the ears. Beginning at about 2300 rpm under part throttle, the exhaust booms like a subwoofer.

Whining about the racket is not meant to subvert the Corvette's expression of speed: Our test car got to 60 mph in just over five seconds, which is a blink slower than the first C5 Corvette roadster we tested last October. It also climbed to 167 mph with the top in place and 160 with the top down. The Corvette roadster growls through the quarter-mile in 13.5 seconds at

107 mph. (Our last hardtop was just two-tenths of a second and 2 mph quicker.) That puts the Vette a half-second ahead of the BMW and 1.2 seconds in front of the Porsche.

The aluminum pushrod V-8 in the roadster makes 345 hp, but more notable in this field of roadsters is the engine's 350 pound-feet of torque, which accelerates the car out of corners incredibly quickly, even at part throttle. You get a sense there's a lot of power in reserve here. The six-speed manual transmission in our test car features a sixth gear so tall that at 70 mph the engine turns over at a mere 1550 rpm. It also features the annoying fuel-saver skip-shift feature that forces you to upshift from first to fourth gear when tooling along at part throttle at slow city speeds.

For our selfish and fuel-consumptive back-road business—conducted on some of the very tight 10Best Roads of the Southeast (*C/D,* January)—we ignored the Corvette's top three gears, saving them for highway cruising. The Corvette launches quickly from corner to corner on the roads that are the most fun to drive. It pulls so strongly you can sometimes avoid downshifting and still maintain as much speed.

It feels balanced on the twisty roads and also in our handling and emergency-lane-change maneuvers. "The Corvette's chassis deserves respect—it's utterly predictable," said senior technical editor Don Schroeder. The Corvette has better grip on the skidpad than the Porsche or BMW, and it outruns both of them through the emergency-lane-change contest.

So why does the car that generates the best numbers in so many categories finish third? In this test, its bulk got in the way, detracting from the complete roadster driving experience.

"The Corvette is just too big here, in reality or perception," said technical editor Larry Webster.

We held this comparison test in South Carolina and set up a handling course on the fast and smooth Laurens proving ground owned by Michelin North America. The fast track should have favored the powerful Corvette, but Mother Nature gave the best time to the mid-engined Porsche in the form of a traction-limiting sprinkle of rain.

"On the track it feels big and brutish, although fast, too," said Schroeder. "On the damp track, though, it can't put power down without extreme oversteer—not like the Porsche, which remains neutral. You find, getting out of the other cars and into the Corvette, that it takes a while to get used to the bulk."

Some staffers would be inclined to purchase the Corvette simply because of its sheer speed—no soft-top car that costs less can run faster. Others think the Corvette's roomy 14-cubic-foot trunk, spacious interior, and amenities make it the best choice of the three for serving as both a weekday commuter and a weekend warrior.

But we felt that the Corvette was happier on a dry test track than it was on these very twisty back roads. As a result, when it comes to delivering the variety of sensations that only come from a topless car, the Corvette loses by a whisker to its smaller, tidier competitors.

2nd Place
BMW M Roadster

We'd drive the M roadster about 60 percent of the time with its top down, for several compelling reasons: The heated seats and the reasonably draft-free cockpit mean you can drive alfresco in winter, your view out of the BMW is vast and expansive, and the top is electrically powered and lowers or rises in 10 seconds. Of the three cars in this group, this is the closest in feel to a motorcycle, and a lot of that sensation is because the doors seem low. Your left shoulder is several inches above the beltline of the driver's door. The

BMW M Roadster

Highs: Drivetrain-refinement perfection, stylish upholstery wardrobe.

Lows: The older-generation rear suspension isn't as flawless as the state-of-the-art M3 sedan's, and you need more time to develop trust in the BMW roadster's handling.

The Verdict: The classic definition of a roadster.

C/D Test Results

	acceleration, seconds							top speed, mph	braking, 70–0 mph, feet
	0–60 mph	0–100 mph	0–120 mph	1/4-mile	street start, 5–60 mph	top gear, 30–50 mph	top gear, 50–70 mph		
BMW M ROADSTER	5.4	13.9	22.2	14.0 @ 100 mph	5.9	6.9	6.7	137 (governed)	172
CHEVROLET CORVETTE CONVERTIBLE	5.1	11.6	16.8	13.5 @ 107 mph	5.7	13.6	12.7	167	172
PORSCHE BOXSTER	6.1	16.8	27.6	14.7 @ 94 mph	7.3	9.8	9.6	146	164
TEST AVERAGE	*5.5*	*14.1*	*22.2*	*14.1 @ 100 mph*	*6.3*	*10.1*	*9.7*	*150*	*169*

Vital Statistics

	price, base/ as tested	engine	SAE net power/torque	transmission/ gear ratios:1/ maximum test speed, mph/ axle ratio:1	curb weight, pounds	weight distribution, % front/rear
BMW M ROADSTER	$43,245/ $43,245	DOHC 24-valve 6-in-line, 192 cu in (3152cc), iron block and aluminum head, Siemens MS 41.1 engine-control system with port fuel injection	240 bhp @ 6000 rpm/ 236 lb-ft @ 3800 rpm	5-speed manual/ 4.21, 2.49, 1.66, 1.24, 1.00/ 36, 60, 90, 121, 137/ 3.23	3080	50.0/50.0
CHEVROLET CORVETTE CONVERTIBLE	$45,619/ $49,235	pushrod 16-valve V-8, 346 cu in (5665cc), aluminum block and heads, GM engine-control system with port fuel injection	345 bhp @ 5600 rpm/ 350 lb-ft @ 4400 rpm	6-speed manual/ 2.66, 1.78, 1.30, 1.00, 0.74, 0.50/ 50, 75, 103, 134, 167, 140/ 3.42	3260	51.5/48.5
PORSCHE BOXSTER	$40,077/ $46,385	DOHC 24-valve flat-6, 151 cu in (2480cc), aluminum block and heads, Bosch DME engine-control system with port fuel injection	201 bhp @ 6000 rpm/ 181 lb-ft @ 4500 rpm	5-speed manual/ 3.50, 2.12, 1.43, 1.03, 0.79/ 34, 56, 83, 116, 146/ 3.89	2900	46.2/53.8

Curiosities

👍 **Rear storage:** The Corvette provides 14 cubic feet of luggage space plus underfloor bins for concealing small items and the CD changer.

👍 **Front trunk:** The Boxster's front cargo compartment (right) is almost big enough to stuff in another flat-six engine.

👍 **Wind blocker:** Only the Boxster has a rear wind blocker, which allows top-down motoring without the usual buffeting.

👍 **Trunklid handle:** The BMW's trunk button is surrounded by a cute chrome flare that makes the lid easy to open, and it looks and feels expensive.

👍👎 **Passenger-airbag switch:** The BMW is the only car with a switch (left) to disarm the passenger-side airbag, but this switch's placement where a gauge belongs is roadster sacrilege.

👍 **Rear cowling:** The hard cover that hides the Corvette's roof when it's down is classy and neat and gives the car a finished look the others lack.

👎 **Emasculated exhaust tips:** The Corvette's power is flawless, but the exhaust pipes (right) seem lost in the large cutout area under the rear bumper.

corner of your left eye picks up the texture of quickly moving pavement close to the car, which brings home the sense of speed you get in this roadster.

And that sense of speed is very real. The M roadster bests its M3-sedan sibling to 60 mph by a 10th of a second, getting there in 5.4 seconds. That's approaching the acceleration of the Corvette, although the big-boy Chevy has *105* more horsepower. The quarter-mile blows by in 14 seconds flat at 100 mph, 1 mph better than the heavier M3 hardtop.

"This motor sings along at 7000 rpm with nary a vibration," said Webster. It's also a flexible powerplant. In the conversion to the 3.2-liter engine, BMW added a freer-breathing dual exhaust system that the M3 doesn't have yet. Output remains at 240 hp and 236 pound-feet, but the torque curve is widened a bit. Add that to the lighter weight of the little roadster—3080 pounds compared with the M3's 3248 pounds—and the overall effect makes the M3 feel slower. BMW admits the four exhaust pipes were added mostly for the macho look, but the next generation of the M3 (due this summer) will likely get the new exhaust system and a power increase. The bigger exhaust doesn't mean the car is noisier, however. At full throttle with the top up, the M roadster is notably more muffled than the two other roadsters. M roadsters destined for the U.S. have less sound-deadening material inside than do European-spec M roadsters, which have been on sale overseas for a year.

The M roadster gets lower-profile front tires and wider rear tires than the Z3 2.8 we tested in our previous roadster comparison in April 1997, but it keeps the same size fenders as the Z3 2.8. Tuning the suspension for these tires required different spring rates and stiffer shock settings, but the goal of this tuning wasn't to make the car quicker on a racetrack, but more civilized around town. It's truly comfortable and easy to drive four-fifths of the way to its limits. Suspension bits start moving around an awful lot when the going gets quicker. Compared with the

roadholding, 300-foot skidpad, g	emergency-lane-change maneuver, mph	idle	full throttle	70-mph cruising	70-mph coasting	EPA city	EPA highway	C/D 600-mile trip
0.88	64.4	50	79	76	75	20	27	17
0.90	64.9	59	85	75	74	18	28	14
0.86	61.3	55	87	73	72	19	26	17
0.88	63.5	55	84	75	74	19	27	16

wheelbase	length	width	height	fuel tank, gallons	front	trunk	front suspension	rear suspension	brakes, front/rear	tires
96.8	158.5	68.5	49.8	13.5	47	5	ind, strut located by a control arm, coil springs, anti-roll bar	ind, semi-trailing arms, coil springs, anti-roll bar	vented disc/ vented disc; anti-lock control	Michelin Pilot SX; F: 225/45ZR-17, R: 245/40ZR-17
104.5	179.7	73.6	47.7	19.1	52	14/11 (top up/down)	ind, unequal-length control arms, transverse plastic leaf spring, anti-roll bar	ind, unequal-length control arms, transverse plastic leaf spring, anti-roll bar	vented disc/ vented disc; anti-lock control	Goodyear Eagle F1 GS EMT; F: P245/45ZR-17, R: P275/40ZR-18
95.2	171.0	70.1	50.8	15.3	47	9	ind, strut located by a control arm, coil springs, anti-roll bar	ind, strut located by 1 trailing link and 2 lateral links, coil springs, anti-roll bar	vented disc/ vented disc; anti-lock control	Bridgestone Potenza S-02; F: 205/55ZR-16, R: 225/50ZR-16

Porsche Boxster

Highs: The intuitive communication through the controls to the driver creates the magic of true partnership.

Lows: Weak motor means it's a dance partner not into full-tilt boogie.

The Verdict: Although many will buy this new Porsche for show, its more secret abilities blossom in private on small, intimate roads.

two other roadsters, the BMW exhibits the most body roll, squat, and dive. On some tight corners, you can lift the inside front wheel off the ground several inches.

"You have to be absolutely precise and smooth through the lane change, or this thing slides around and is slow. Way twitchier than the Porsche," said Webster.

"It's the least composed of the three roadsters on the damp track," said Schroeder. "Steering requires frequent correction, and you must compensate for weight transfers." What he means is that lovers of the older-generation Porsche 911 who liked driving sideways can reminisce in this BMW.

The BMW's leather seating surfaces and upholstery are beautifully color-contrasted, and the gauges get chrome bezels. Inside, this car is the most attractive of the three in this test. Outside, brake scoops replace the fog lights in front, and the rear license plate moves up from the bumper to the trunklid. We like it a lot. The Corvette costs $6000 more as tested, and it boasts a string of equipment unavailable on the BMW, such as dual climate controls, run-flat tires with integral pressure-sensor gauges, a power antenna, extra 12-volt outlets, a compact-disc changer, memory seats, and more. Those who have come to expect all of this on a luxury-priced two-seater will miss it on the BMW.

1st Place
Porsche Boxster

Not only would we take this car to the ski mountains in January with the top down, but we'd have to force ourselves to consider the downside of the latest Hong Kong flu before we'd put the top up, even when it rains in the summer. That means we'd want to drive this car topless about 80 percent of the time we spent in it. Maybe we're dreaming, but just in case, we'd pack one of the Boxster's two spacious trunks with extra jackets for the passenger.

It's not uncomfortable in wintertime, at least a South Carolina winter, and we're not exaggerating here: The Boxster's high beltline means your shoulders ride level with the tops of the doors, so when the side windows are up, they block a lot of wind. Perforated panels covering the insides of the roll hoops and a plexiglass barrier between the seats also keep the air still way down into the footwells of the car. All of this aids the heater's ability with the top off. The Boxster is the most intimate-feeling of these three roadsters.

Successful top-down climate management is just one reason the Porsche is a good roadster. Agility, involvement, feedback, balance, sensitivity, comfort, and refined behavior in a variety of conditions are the others.

At first, the Boxster feels less powerful than the two other roadsters here. It runs the quarter-mile more than a second behind the Corvette (at 14.7 seconds). Its 7.3-second rolling-start acceleration from 5 to 60 mph feels positively sluggish after you've driven the two other screamers, both of which manage the task in less than six seconds.

Top speed is ungoverned at 146 mph—that's about as fast as a V-8 Mustang. At that speed, the Boxster feels stable and doesn't get blown around by sidewinds. Of course, the rocket Corvette goes 21 mph faster.

While you're indulging in other full-throttle behavior, the Porsche feels quiet and confident. The two other roadsters seem to yell and shout. The Boxster is quieter than the others when cruising with the top up, and depending on how high you hold your head, it's a whole lot quieter when the top is down.

Roadholding ability is less than the two other roadsters' at 0.86 g, and the Porsche was at least 3 mph slower than the others in the lane-change test. Yet the Boxster managed to run this test without leaving a mark on the asphalt. The first time through our cone-marked course, the BMW left long, wide, dark tire smears of cooked rubber—four of them. The Corvette, too, autographed the pavement and made screeching and wailing noises at its limit. The Porsche remained unruffled and unprovoked.

Out on real roads the Boxster proves its mettle. While churning the steering wheels of all three cars back-to-back on no fewer than six of the roads we divulged in January as the 10Best Roads of the Southeast, the Porsche never fell behind, despite the enormous difference in acceleration times.

"Everything is so direct, quick, and immediate in this car—the chassis, the steering response, the roll control. It feels like the most nimble car here, which makes up for the lack of horsepower a bit," justified Schroeder.

"I think the M roadster is more fun, but the Boxster is pretty damn fun without scaring you. It's a tough call which is better," concluded Webster.

It may be a tough call, but it's one we've made twice now. In our comparison test last April, the Boxster earned almost exactly the same ratings that it received during this test, even though the tests occurred 2800 miles and 11 months apart. The scores for the early test's Z3 2.8 and SLK230 were lower than those earned by this test's Corvette and M roadster, which tells us these two latest roadsters are closer to our ideal. And our conclusion, once again, is that this Porsche is less handicapped by its moderately powered engine than you'd think. We can't wait for the arrival of the rumored 250-hp Boxster S model this summer, with power enough for the truly impatient. •

Editors' Ratings

	engine	transmission	brakes	handling	ride	driver comfort	ergonomics	features and amenities	fit and finish	value	styling	fun to drive	OVERALL RATING*
BMW M ROADSTER	10	10	10	8	8	8	9	4	9	8	10	10	94
CHEVROLET CORVETTE CONVERTIBLE	10	9	10	10	9	10	9	10	8	9	9	9	93
PORSCHE BOXSTER	8	9	10	10	8	9	9	7	9	8	9	9	95

HOW IT WORKS: Editors rate vehicles from 1 to 10 (10 being best) in each category, then scores are collected and averaged, resulting in the numbers shown above.
* The overall rating is not the total of those numbers. Rather, it is an independent judgment (on a 1-to-100 scale) that includes other factors–even personal preferences–not easily categorized.

BMW M Roadster

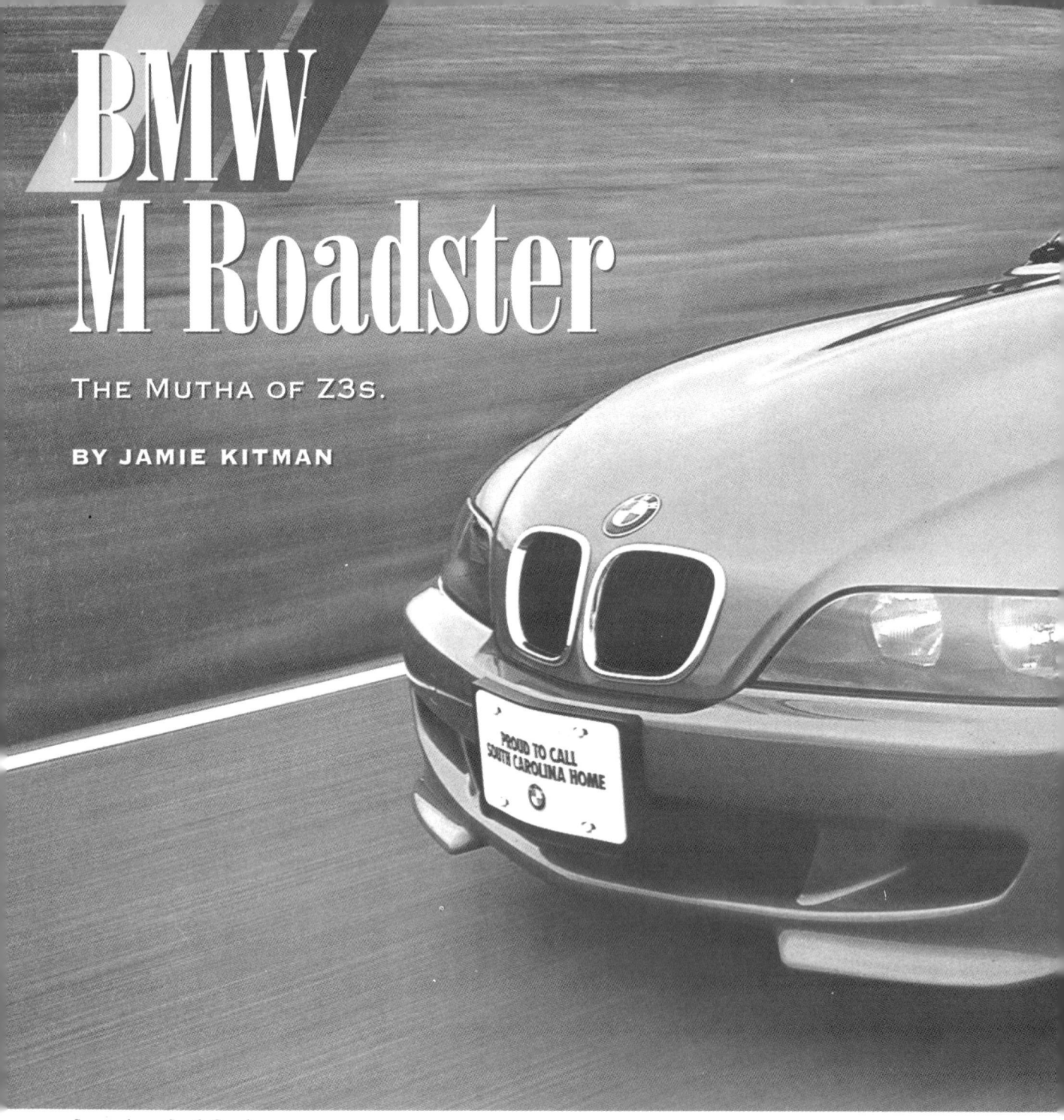

THE MUTHA OF Z3s.

BY JAMIE KITMAN

Spartanburg, South Carolina—

It's murder out there. New, quality sports cars enter the field of mortal combat seemingly every week. So it's a good thing BMW had an M version of its successful Z3 roadster waiting on the bench. Here in BMW's adopted no-nonsense home, the land of the fighting Gamecocks, the language of the gridiron can explain why.

The new M roadster assembled here is another niche BMW that helps to define a market segment—near supercar. A team player that carries the ball usefully downfield, it offers BMW fans their money's worth, and more, while buying the good ol' boys from Bavaria yet another first down. It also happens to be the fastest BMW ever officially sold in America, if not the best.

In BMW argot, M stands for Motorsport, but to you and me, M stands for More. Twenty-six years of delectable exercises in high performance have come and gone since the first 3.0CSL, and what we've grown to expect from M cars is mostly more power, more handling, and more brakes.

Of course, in line with the industry's growing infatuation with sub-branding and upscale micro-marketing, there are more M variants now than ever before, and BMW will sell more of them than ever, too.

That's one reason M roadsters are built on the assembly line alongside regular production cars. Thankfully, there hasn't been any standard BMW less than mighty fine for a long time. And it's hard to imagine an M edition that's ever really disappointed.

So whaddya know? This one doesn't, either.

What makes a roadster M? Start with the 3.2-liter DOHC 24-valve

PHOTOGRAPHY BY GLENN PAULINA

in-line six out of the M3, developing an identical 240 bhp and 236 pound-feet of torque. Delete the automatic transmission option. Install the M3's mondo brakes (fully vented, 12.4 inches front and 12.3 inches rear) and some jumbo rubber (225/45ZR-17 front, 245/40ZR-17 rear) riding on king-size alloy wheels (17 x 7.5 inches in front, 17 x 9.0 inches at the rear).

Different front and rear aprons, quad exhaust pipes, and chrome side-grille trim distinguish the M roadster further, along with the availability of two exclusive colors (Imola red and evergreen). Inside, color-keyed sport seats and steering wheel, chrome instrument surrounds, and an illuminated shift knob are key emblems of M-ness, as is a standard power roof.

Behind the roadster's new mask and beneath its flamboyant flares lie the M3's variable-assist, variable-ratio power steering and a tweaked suspension. Ride height is dropped an inch, firmed-up dampers are selected, and the front suspension geometry is modified, à la M3, with a resized anti-roll bar linked to the struts by way of ball joints. Around back, fortified semi-trailing arms and a reinforced rear crossmember carry the load.

The limited-slip differential seen in the Z3 2.8 remains on duty, but, surprisingly in these litigious times, traction control is not available. According to M brand manager Erik Wensberg, this apparent oversight is because BMW engineers have yet to perfect a system that can deal with the exigencies of the steroidal roadster's compact wheelbase. (But traction control is standard on other Z3 models.)

"Be careful!" Wensberg repeatedly cautioned a passel of lead-foot journalists as we set out on a wet and intermittently freezing morning for a day of M roadster appreciation along a potpourri of potentially deadly hill-country back roads. Patches of snow, ice, and sand; narrow tree-lined passes; and precipitous drop-offs did their part to raise the specter of eternal darkness. The good news was, if we survived, we'd get another chance to meet our maker the following day at the track.

Wensberg's introductory assertion that the M roadster's performance "will rival even the greatest sports car legends such as the AC Cobra" didn't help steady the press corps' nerves. A tail-happy BMW with the ability to rip off 5.5-second rushes to sixty could in theory tread in the Cobra's treacherous footsteps.

But it didn't. We've met the Cobra (well, a pretty convincing replica), and the M is no Cobra. It is seriously rapid, yes. Given the right set of circumstances and a series of erroneous operator inputs (fewer still if you're a real idiot), you could probably even kill yourself in it.

And we all know short wheelbases and gobs of power can be recipes for disaster. But the M roadster is no widow-maker. It wants to go fast; it doesn't want to bite. Lots of grip, as you'd expect from those wide tires, and then it's time for opposite-lock therapy. Under the circumstances, the M roadster rides pretty well, too.

In short, here is a hale and hairy sports car, easily the most exciting of the Z3 variants, with handling that invites—rather than discourages—mass exploitation of 240 innocent horsepower.

Still, we can't help feeling the M roadster is something less than the ultimate ultimate driving machine. The previous four-cylinder M3, which did sterling work with a very similar rear suspension layout, always felt more alive. That doesn't mean the M roadster goes to the Hall of Shame. Along with the upcoming M coupe, this diminutive rocket ship carries the ball deep into enemy territory. It's not a supercar, but damn close. Prospective Chevrolet Corvette convertible buyers and Porsche Boxster S intenders may want to think twice.

Chrome bezels for instruments and controls, a revised center console design, special sport seats, a two-tone interior color scheme, and numerous exterior styling cues such as chrome side grilles and redesigned side mirrors enhance the aesthetic appeal of the BMW M roadster. From the rear, the nine-inch-wide rims and four-tipped exhaust hint that this is no four-banger.

BMW M ROADSTER
Front-engine, rear-wheel-drive convertible
2-passenger, 2-door steel body
Base price (estimated) $42,000 (+ luxury tax of 7% over $36,000)

ENGINE:
24-valve DOHC 6-in-line, iron block, aluminum head
Bore x stroke 3.40 x 3.53 in (86.4 x 89.6 mm)
Displacement 192 cu in (3152 cc)
Compression ratio 10.5:1
Fuel system sequential multipoint injection
Power SAE net 240 bhp @ 6000 rpm
Torque SAE net 236 lb-ft @ 3800 rpm
Redline 7000 rpm

DRIVETRAIN:
5-speed manual transmission
Gear ratios (I) 4.21 (II) 2.49 (III) 1.66 (IV) 1.24 (V) 1.00
Final-drive ratio 3.23:1

MEASUREMENTS:
Wheelbase 96.8 in
Track front/rear 55.0/58.7 in
Length x width x height 158.5 x 68.5 x 49.8 in
Curb weight 3084 lb
Weight distribution front/rear 51/49%
Coefficient of drag 0.42
Fuel capacity 13.5 gal
Cargo capacity 5.1 cu ft

SUSPENSION:
Independent front, with damper struts, lower control arms, coil springs, anti-roll bar
Independent rear, with dampers, semi-trailing arms, coil springs, anti-roll bar

STEERING:
Rack-and-pinion, variable-ratio, variable-power-assisted
Turns lock to lock 3.2
Turning circle 34.1 ft

BRAKES:
Vented discs front and rear
Anti-lock system

WHEELS AND TIRES:
17 x 7.5-in front, 17 x 9.0-in rear cast aluminum wheels
225/45ZR-17 front, 245/40ZR-17 rear Michelin Pilot SX MXX3 tires

PERFORMANCE (manufacturer's data):
0–60 mph in 5.5 sec
Top speed (electronically limited) 137 mph
Pounds per bhp 12.8
EPA city driving 20 mpg

BMW Z3

The BMW Z3 is road candy; one test drive is all the definition you'll ever need. It's a roadster, plain and simple, a 2-seater with a single-layer fabric top. Standard power windows are a concession to modernity, but the canvas goes up and down by the Jack Armstrong method—although it's actually easy to drop or lift—and it attaches to the windshield header with conventional manual latches.

Of course, the Z3 is more than just a roadster. It's a roadster by BMW, and it goes, handles and stops like a BMW. Its unitized steel body/chassis is Z3 specific, though it borrows its front suspension (MacPherson struts and arc-shaped lower arms) from the 3-Series BMWs. At the rear, the Z3 shares its semi-trailing-arm rear suspension with the 318ti. The base 4-cylinder engine dittos the 138-bhp twincam four in the 318, and so equipped the roadster is called the Z3 1.9.

Drop in the aluminum-block 2.8-liter inline-6 and output climbs to 189 bhp, with torque a rippling 203 lb-ft at 3,950 rpm. But the Z3 2.8 is more than an engine swap. It includes a heavier-duty Getrag 5-speed (optional automatic with either engine), plus stronger rear suspension with a wider rear track, revised suspension geometry and calibration, and bigger brakes. Z3 spotters note the wider rear flanks, unique front bumper and spoiler package and special wheels.

Either way, the Z3 is strictly for two, with little room to stash miscellanea behind the seats. No surprise, with a wheelbase 10 in. shorter than the 3-Series models. The trunk is roomy enough, though, to swallow a soft golf bag.

The cockpit's dual-cowl style is reminiscent of old English roadsters, adding modern touches including dual cupholders, the option of Harmon/Kardon sound (standard on the 2.8) and a CD player. Body styling is all BMW with a twin-kidney grille and fender vents recalling the classic BMW 507 of the '50s.

New on '98 Z3s is a standard rollover bar and the option of sport seats in the 1.9 and 2.8. A Premium Package, with leather, wood and a power top, can come on either model as well.

Future treats include an "M" version with 240 bhp, and an MGB GT-like Z3 coupe. Just check the candy counter at your BMW store.

QUIKFACTS

MODEL	MSRP	ENGINE	TRANS	ABS	SEATS	A/C
Z3 1.9	$29,425	I-4	5M	STD	2	STD
Z3 2.8	$37,900	I-6	5M	STD	2	STD

SPECIFICATIONS

Layout	rwd
Wheelbase	96.3 in.
Track, f/r	55.6/56.3 in.[1]
Length	158.5 in.
Width	66.6 in.[2]
Height	50.7 in.[3]
Curb weight	2690 lb
Base engine	138-bhp dohc 16V I-4
Bore x stroke	85.0 x 83.5 mm
Displacement	1895 cc
Compression ratio	10.0:1
Horsepower, SAE net	138 bhp @ 6000 rpm
Torque	133 lb-ft @ 4300 rpm
Fuel econ, city/hwy	23/32 mpg
Optional engine(s)	189-bhp dohc 24V I-6
Transmission	5M, 4A
Suspension, f/r	ind/ind
Brakes, f/r	disc/disc, ABS
Tires	225/50ZR-16
Luggage capacity	5.0 cu ft
Fuel capacity	13.5 gal.
Warranty, years/miles:	
Bumper-to-bumper	4/50,000
Powertrain	4/50,000
Rust-through	6/unlimited

[1] Z3 2.8 55.6/58.8 in.
[2] Z3 2.8 68.5 in.
[3] Z3 2.8 50.9 in.

MACHISMO

Story: Jeremy Clarkson

The **BMW Z3** is nice enough, but it's a bit of a **girlie's car**. Isn't it? **Not** any more. The **M Roadster** version has arrived in a cloud of **testosterone** and **tyre smoke** to stomp all over that **effete** image. But **is it man** enough to take on **Britain's** own chestwigged **charmer**, the **TVR Chimaera**? We took the **soft**-topped **hard** chargers to a suitably **gritty** airfield and let our own **models** of **masculinity**, **Jeremy** and **Tiff**, strut around in them, **pose** in them, and even **drive** them in a suitably **blokey** fashion with loads of gratuitous **wheelspin** and a lot of **pointless** revving of the engines. So did the German **thugmobile** thrash our **ladwagon**? Read on, if you **can**

At last – a Z3 that will do squealy sideways stuff with the right amount of enthusiasm. Unlike Jeremy, who is still convinced that the BMW is only fit for footballers' wives and off-duty estate agents. He's warming to it, though…

With his chiselled jaw line and boyish haircut, Tiff is one of the most charming and handsome people you could ever wish to meet. He's funny, personable, unaffected by fame, and a truly great driver.

The only trouble is that, when it comes to sports cars, Tiff talks out of his bottom. If you were to give him a cardboard box with a multivalve engine and rear-wheel drive he'd love it.

So, when BMW announced that its horrible little hairdryer, the Z3, was to be fitted with the 321bhp, 24-valve, straight-six M engine, Tiff was like a man with two willies.

I'm rather different. While I like a car to turn in with verve, I also like to have some Verve on the stereo. I understand that half the fun of a sports car is in tooling down the Champs Elysées in summer, enjoying the style and the noise. I'm more of a TVR man frankly.

So, at Abingdon Airfield, one bitterly cold morning, the scene was set for a battle royal, as Tiff and I tried to make the other see sense.

He'd brought along his M Roadster, hoping to show me just how fast it could go, and I had a Chimaera, with a new and huge five-litre V8, eager to prove that the intoxicating noise easily

on test

compensates for half a g in the corners.

Some would see this whole exercise as pointless. I mean, you're either a TVR man or a Bee Em man and ne'er the twain shall meet. And on top of that, both are 320bhp, front-engined, rear-drive, drop-top two-seaters… in the same way that Tony Blair and Alan Clark are both politicians. Believe me, these cars are very different indeed.

The TVR looks bulky and aggressive but thanks to a plastic body, it's significantly lighter. Its spacious cockpit is a dream too, with wood, leather and polished aluminium styled by someone who truly understands interior design. The heater, for instance, is the sort of touch that Ridley Scott would use in a blockbuster sci-fi film.

Then you've got the heavyweight and really rather unattractive BMW which, inside, is simply terrible. There's acres of plastic-look leather in two shades of blue which, in this M version, has been enlivened with a welter of chrome-ringed analogue dials. It just doesn't work, in the same way that antiques don't work in a Bryant starter home.

But the worst thing about the Bee Em is its lack of space. Even Tiff, who doesn't quite make the six-foot mark, struggled to get his legs under that barge-large steering wheel. Further back, lack of room is an even bigger problem. While the TVR has a boot

Tiff is a man of sophisticated and well-honed tastes when it comes to cars; the brutal TVRs have never been his favourite things. But actually, when you come down to it, ludicrous power oversteer is rather enjoyable after all…

It might come from Blackpool, the chip butty capital of Britain, but the Chimaera is a sleek and sexy thing. The curvy interior matches the radical exterior, and the knob between the seats matches the one behind the wheel. Oh, that's Jeremy. He never did quite get the idea of the hood

you could call generous, the BMW comes with something that won't double as a handbag. You don't get a spare wheel; not even a space saver.

But you do get an electric hood, whose motor robs even more precious boot space. Mind you, the centre bit of the TVR's roof has to go in its trunk, so let's be fair; let's say neither is a rival for the Renault Espace.

Overall, though, on style and practicality the TVR romps into an early lead which it sustains by being better value. The M Roadster is £40,570 while the TVR is just £35,850 plus an essential £980 for power steering.

Fuel consumption doesn't matter with cars of this type, but depreciation does. And as a betting man, I'd say they'd be about evens. That said, the TVR may be more costly to run in terms of heartache. You are very likely to be beset with leaks and electrical malfunctions which will drive the faint-hearted mad. I know of one girl whose Chimaera is now growing mushrooms in the boot and on our test, an electric window temporarily packed up. Mind you, the BMW's power steering started to make odd whining noises too, but that was after Tiff had performed 416 doughnuts.

But now it's time for some proper driving, time to show Tiff that his BMW is a jumped-up kitchen utensil. I knew I could bring him round... but there was a surprise in store.

Getting the TVR off the line quickly is a nightmare. You need to crawl away, feeding in the power so gently that people begin to think Quentin Willson is at the wheel but then, just when you're starting to think a bus would be faster, you plant your foot hard down, grab second and flash past 60 in just over five seconds.

The BMW doesn't have traction control either but it does wear 245 section rear tyres on nine-inch wheels so, in theory, you have more grip for lightning starts. Even so, it went three tenths slower to 60 than the TVR.

In the midrange, the TVR continues to defy logic. The BMW has an exceptionally torquey motor but still it can't keep up with the Chimaera. Any revs, any gear, and the TVR just blankets its opponent in a wall of noise and scuttles off in a relentless quest to headbutt the horizon. But the TVR doesn't have it easy. The BMW's never far behind and superior aerodynamics see it draw level at 120mph and then ease ahead.

But then you go ballistic. Long after the busybody limiter has called a halt to proceedings at 155mph in the M,

84

on test

the Chimaera is still going; all the way, TVR says, to a flat-out max of 167mph.

More than this, the TVR *feels* so much faster. Put your foot down and, amid the crescendo of noise, the nose reaches for the stars, the back steps out of line and you're off, like you're riding a thunderstorm.

The BMW isn't what you'd call quiet but there's no pitching back and forth and all remains serene as you flash from 40 to 120mph. The TVR may be faster but this is much more relaxing. The M also stops more quickly thanks to its compound, split hub, vented, cross-drilled race discs with ABS. The TVR just has brakes.

And then there's the rest of the underside. Now a TVR is basically a box girder bridge whereas, you're thinking, this little Roadster has come from BMW's Motorsport department. Yes, but don't be fooled here.

It may have the same 321bhp engine as the awesome M3 Evo but there was no room to fit the new six-speed box, so it makes do with the old five-speeder. And instead of the Z axle, you get a conventional set up with stiffer bushes just like the original four-cylinder M3. It's not quite the technocrat those M badges might have you believe.

So, as Tiff and I thundered toward the first corner, we knew the results would be interesting.

Both, as you'd expect, are set up to understeer at first, but both will break traction with consummate ease if, halfway through the bend, you either lift off or brutalise the throttle. Then things get very different indeed.

In the BMW the back slides round, nice and gently, giving you plenty of time to balance the throttle, apply some opposite lock and pull a cool face for the cameras.

In the TVR, the back slides round, and keeps on sliding round. You can pull whatever face you like and it won't matter because you'll be going along backwards and no-one will see. I tried, time and again, to get the Chimaera to sustain a power slide and on each occasion it ended in tears.

Happily, I'm old enough now to accept that it requires a little more in the way of talent than I possess, so I handed the TVR to Tiff who, though better, still had a hell of a job keeping the TVR going vaguely in his chosen direction. The body was pitching and rolling all over the place.

I decided that in a TVR, which has less grip and is harder to control, it's better to pussyfoot round the corners and use your superior firepower on the

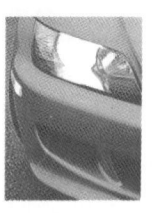

The BMW is a neat updating of the retro theme; it has little chrome bits all over the place. The interior, though, is in questionable taste even if the windscreen keeps almost all the wind away. Either that, or Tiff's coiffure is actually a carbon fibre and Kevlar crash hat in disguise

85

straights. In the BMW, you can go quickly everywhere.

I could only apply the necessary opposite lock if I tucked my left foot under the clutch pedal, otherwise my knees were always in the way of my twirling hands. The steering column doesn't adjust so it's a bloody nightmare, that car, if you're as tall as me.

Even so, it would be my choice. Tiff had convinced me that its supposedly superior build quality and greater agility is more important than the sheer leeriness of the TVR.

Dammit, the man was right. I would have to concede defeat and hand the laurels to the top-of-the-range Z3, a car whose tone, I've always felt, was set by the astonishingly gutless 1.9.

I sauntered over to break the news to Tiff… who I found in a deeply crestfallen state, admitting that he had been convinced by my arguments. The TVR, he admitted, was just too exciting to be overshadowed by such an anodyne car as the BMW.

Shit. I've never done this in any test, ever before, but in this case, it really is a draw. We can't agree on a winner, so I guess they both lose. One of these days, someone is going to make the perfect two-seater sports car – one that growls like a TVR, has the practicality of a Merc SL and the agility of BMW's M Roadster. But it hasn't happened yet and as a result, Tiff and I went home in our comfy saloons □

on test

MACHISMO & GIZMOS

	BMW M Roadster	TVR Chimaera 5.0
Performance		
0-30mph (secs)	2.2	2.0
0-40mph (secs)	3.3	3.0
0-50mph (secs)	4.4	4.0
0-60mph (secs)	5.5	5.2
0-70mph (secs)	7.3	7.1
0-80mph (secs)	9.0	8.6
0-90mph (secs)	11.1	10.6
0-100mph (secs)	13.8	13.4
0-110mph (secs)	17.2	16.3
0-120mph (secs)	21.0	21.0
0-130mph (secs)	26.3	28.0
Max speed (claimed)	155mph (limited)	167mph
Standing qtr/terminal speed	14.1secs/101.1mph	14.0secs/n/a
30-50mph in 3rd (secs)	3.2	2.8
30-50mph in 4th (secs)	4.8	3.9
50-70mph in 5th (secs)	6.1	5.2
30-70mph thru' gears (secs)	5.1	5.0
Braking, 70-0mph (ft)	157.3	173.8
Costs		
On the road price	£40,570	£35,850
Test/combined mpg	17.5/25.4	18 approx/n/a
Insurance group	20	20
Service interval	8,000 miles approx	every 6,000 miles
Warranty	3 years unlimited	1 year unlimited
Equipment		
Driver/passenger airbags	yes/yes	no/no
Air conditioning	option	option
Alarm/immobiliser	yes/yes	yes/yes
ABS/power steering	yes/yes	no/option
Central locking/remote	yes/yes	yes/yes
Electric windows	yes	yes
Electric roof/seats	yes/yes	no/no
Full leather/heated seats	yes/yes	option/option
Limited slip diff.	yes	yes
Radio cassette/CD	yes/option	yes/option
Traction control	no	no
Technical		
Engine	inline 6cyl, 24v, dohc	90deg V8, 16v, pushrod
Capacity (cc)	3,201	4,988
Max power (bhp @ rpm)	321 @ 7,400	320 @ 5,500
Max torque (lb ft @ rpm)	258 @ 3,250	320 @ 4,000
Transmission	5sp manual, rwd	5sp manual, rwd
Brakes (f&r)	floating, vented discs	vented discs
Front suspension	MacP strut	double w/bone
Rear suspension	semi-trailing arm	double w/bone
Wheels	7(f), 9(r)x17" alloy	7x15"(f), 7.5x16"(r) alloy
Tyres	225/45(f), 245/40(r)ZR17	205/60ZR15(f), 225/55ZR16(r)
Weight (kgs)	1,425	1,060
Fuel capacity (litres)	51	57
Dimensions L/W (mm)	4,025/1,850	4,013/1,867

TIFF GOES MUCHO MACHO

I reckon Jeremy has always had pretty basic tastes when it comes to choosing his motor cars. As far as I can see, he's rather hooked on big, brash, noisy motors. Let's face it, they suit his style.

It's the same with sports cars – if it doesn't growl it's not worth knowing. So he's found his perfect partner in the TVR.

So to try and persuade him that the delectable BMW M Roadster – a car that, as far as I'm concerned, has at last elevated the Z3 sports car into the big boys' toys division – was a far more rewarding machine than any of his brutal Blackpool bruisers would obviously be a difficult task.

I knew as I handed the car over to him that he would be in a bad mood before he even got going because the cockpit dimensions suit only those, like me, that are less than six foot tall (though even I was far from happy with the fact that my golf clubs wouldn't fit in the boot).

But, as far as I'm concerned, all that can be forgiven for the sheer joy of flinging the car around our airfield location. Like all BMWs the steering feedback and response are like no other. The lightness of the controls makes every operation a natural move. It's a car that you can make dance to your every wish.

By contrast, after handing the BMW to Jeremy, I was left to play in the Chimaera which seems to fight you with every move. The rear suspension seems to me to have been set up too soft, probably to maximise tyre grip, and as a result it's not so easy to control in a slide as the wonderful BMW. The essential power steering lacks subtle communication, too, and the gearchange is slightly awkward but oh does it have some grunt!

It feels like the TVR has twice the power of the Bee Em and I'm sitting low down in a perfect driving position only spoiled by the lack of room for a left footrest. The interior dials and gizmos with their peculiar ways have an endearing attraction and I can even fit *two* sets of golf clubs in the boot.

And it also makes the most terrific V8 noise. Oh my God, I'm turning in to Clarkson. I'm starting to like this. Look out, here he comes now. He's bound to have hated the Roadster and now I'm going to have to tell him…

BMW Z3 1.9i
Price £21,480
Date tested 26.2.97

Now that we've driven all the Z3 models – right up to the fire-breathing 321bhp M-Roadster – it's easier to put this entry-level 140bhp 1.9 into perspective. No revelations here, we're afraid. This is the weakest and least satisfying Z3 you can buy. It simply doesn't have enough power to express itself; the grip-to-grunt ratio is all wrong for a sports car.

That said, the beefier six-cylinder models haven't exactly turned out to be shining examples of the genre, either.

Z3 1.9 requires severe provocation to elicit this kind of tail out attitude

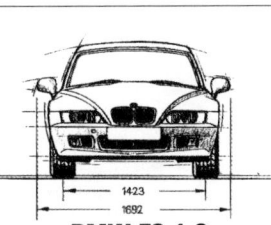

The M-Roadster comes on like a latter-day AC Cobra but can't even slay a TVR, never mind a Porsche 911. The 2.8 offers the best balance of go, grace and value, but is still disdainfully seen off by the Boxster.

So, given that there probably isn't a great car in the Z3 trying to get out, it's easier to like the smallest-engined model. According to BMW's research, only a lowly four per cent of potential Z3 customers are concerned with the way this car performs and drives. Style trumps dynamics on the wish list and, disjointed as its shape is front-to-back, the Z3 has plenty of that.

It's also well made (and getting better as time goes by), comfortable, practical, pretty well equipped and, above all, a BMW. Not a great BMW but a head-turner nonetheless.

BMW Z3 1.9

HOW MUCH?
Price	£21,480
Test date	26.2.97

HOW FAST?
0-30mph	2.7sec
0-60mph	8.4sec
0-100mph	25.8sec
30-70mph	8.2sec
Stand qtr mile	16.3sec/85mph
Standing kilo	29.7sec/104mph
30-50mph in 4th	8.0sec
50-70mph in 5th	11.1sec
Top speed	123mph
60-0mph	2.7sec
Noise at 70mph	76dB(A)
Test weight	1164kg

HOW THIRSTY?
Overall test average	29.3mpg
Touring route	34.1mpg

ENGINE
Layout 4 cyls in line, 1895cc
Max power 140bhp at 6000rpm
Max torque 133lb ft at 4300rpm
Specific output 74bhp/litre
Power to weight 120bhp/tonne

GEARBOX
Type 5-speed manual
Top gear 20.6mph/1000rpm

VERDICT ★★★
Not a true sports car, but as a stylish, well built and surprisingly affordable cruiser it still cuts the mustard.

Z3 borrows switches from Compact

Four-cylinder 1.9 engine lacks grunt

BMW Z3 2.8
Price £28,120
Date tested 3.9.97

BMW's American-built roadster needs the company's sublime 2.8-litre straight six to be an honest roadster – something the four-cylinder 1.9 patently isn't. Unfortunately, it would need intervention from above to be a great one. Not even the M-Roadster's 326bhp is as exciting as it should be.

The 2.8's 192bhp nevertheless turns in a fine set of figures, a 0-60mph time of 6.7sec lagging just two-tenths behind Porsche's Boxster. Moreover, the Z3 actually beats its adversary for

2.8-litre Z3 may be quicker than Boxster but it's nowhere near as fun

in-gear punch. Value isn't in question, either; as well as decent performance, it gets a long list of equipment that includes leather trim, power seats and traction control.

But the great engine and classic rear-drive layout never really come together as you expect them to; certainly not as the retro styling cues suggest they might. The Z3 is based on the humble 3-series Compact and it feels it. The body structure wobbles and shimmies on poor roads, the more so as the broad, grippy tyres bite into the tarmac and increase the suspension loadings.

Tail-out indulgence is possible, but only in low gear bends or in the wet. An MX-5 like balancing act simply isn't in the BMW's repertoire. Probably just as well; it doesn't have the finesse to carry it off.

BMW Z3 2.8

HOW MUCH?
Price	£28,120
Test date	3.9.97

HOW FAST?
0-30mph	2.4sec
0-60mph	6.7sec
0-100mph	18.0sec
30-70mph	5.9sec
Stand qtr mile	15.2sec/93mph
Standing kilo	27.5sec/114mph
30-50mph in 4th	6.1sec
50-70mph in top	8.3sec
Top speed	134mph
60-0mph	2.6sec
Noise at 70mph	na
Test weight	1335kg

HOW THIRSTY?
Overall test average	24.9mpg
Touring route	30mpg

ENGINE
Layout 6 cyls in line, 2793cc
Max power 192bhp at 5300rpm
Max torque 202lb ft at 3950rpm
Specific output 68bhp/litre
Power to weight 143bhp/tonne

GEARBOX
Type 5-speed manual
Top gear 22.7mph/1000rpm

VERDICT ★★★★
Probably the best Z3 in the range. Quick, refined, comfortable and comparatively good value, but it still falls short on fun.

BMW 193bhp straight six still superb

Chrome details give a lift to cabin

SCARY SPICE vs PO

SHOOTOUT | **SPORTS COUPES** — Mid-life crisis? Then spice up your life with BMW's brutal new

Time for an *Autocar* mid-life crisis. Porsche 911s, we've had them all – fast, practical, the everyday supercar. But the latest 996? Too much Grand Turismo and not enough sports car to replace the previous 993 – that's the rumour. And the same problem applies to the Jaguar XK8, even the XKR. Ferrari F355? Good question, great car. But it costs a fortune.

Doesn't leave much real choice, if you want an instantly recognisable, prestigious name tag on a new two-seat sports coupe. Mercedes-Benz? Nobody could seriously suggest the SLK is in the major performance league, and the SL's ready to be superannuated. No, what we want is the look and feel of sophisticated engineering, and performance that instantly revitalises the most crisis-bound, middle-aged blood in one lusting rush to the red line.

BMW? No, not the toy-like Z3, but the new Z3 coupe. More practical and comfortable than the roadster and just as exciting. Is it more of a sports car, more able to satisfy middle-aged yearnings, than the new 911? Let's find out.

Parked together, the disparity could hardly be more telling. One, obviously, was designed, carefully honed to appeal to a mature customer. The other, equally obviously, seems the figment of one maniacal man's wonderfully demented imagination. The 911 is graceful, cohesive, a design that successfully integrates the Porsche tradition in a new longer and wider shape that's both pragmatic and elegant, despite the obvious commonality with the Boxster

BMW wheels flashy, like rest of car

Stock 17in alloys suit 911 superbly

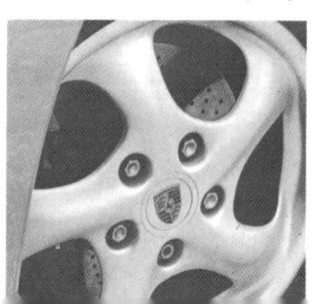

RSCHE SPICE

Z3 M coupe or Porsche's 911 Carrera. By Peter Robinson

Lights are least radical part of Z3

Every component spells Porsche

BMW straight-six has had M boost

Boxer six sits behind rear wheels

forward of the B-pillar.

Doesn't the M coupe look much smaller – it's a significant 405mm shorter – and wilder? Talk about road presence. Few contemporary cars demand your attention like the M coupe. But it's not really a coupe at all. More a sports estate, a genre that's given a kick-start every decade or so, flourishes through its 15 minutes of fame and then quietly dies away.

Will the M coupe go the same way as the Volvo ES1800 and the Reliant Scimitar GTE? Not if BMW can help it.

BMW took the basic layout of the Z3 in its most formidable, M-powered form – and that, as any enthusiast knows, means 321bhp developed at a breathtaking 7400rpm – added a long, fixed steel roof and an opening tailgate. Hot hatch? Yes, sort of, except in this case the engine is mounted up front longitudinally, and drives the rear wheels through a five-speed gearbox, because there's no room for the six-speed 'box now fitted to the M3.

Still, there's no electronic gimmicky to breach a direct relationship between driver's right foot and the massive 245/40 ZR17 rear rubber. BMW is rightly proud of its engines and the purposeful in-line six resides under a massive bonnet that's virtually the entire nose of the Z3.

With those proportions the M coupe couldn't be anything but front engine, rear drive. By siting the cabin, and especially the two occupants, in the rear half of the 2459mm wheelbase and locating items like the battery in the floor of the luggage area, BMW has been able to retain its customary 50/50 weight distribution.

In Porsche's 50-year history the standard engineering practice is a horizontally opposed engine mounted in the rear. The 911's all-new unit is now water-cooled, of course, but it's still stuck out beyond the rear axle line, so a seemingly unfashionable 61 per cent of the car's weight hovers over the rear wheels, meaning it remains tail heavy. Not that you see much of the engine under the tiny rear-opening grille.

Water-cooling allows Porsche to use four valves per cylinder and quad cams. But even so, and despite an extra 186cc, it can't quite match the output of the BMW. The M coupe achieves 100bhp per litre, the Porsche 87bhp per litre for 296bhp at 6800rpm, with a 7300rpm red line, just 100rpm shy of the long-stroke BMW.

Despite the disparity in their layouts, both engines pump out an identical 258lb ft of torque, though the BMW peaks at 3250rpm, while the Porsche's is at a rather higher 4600rpm.

The Porsche has a number of small advantages. In isolation they might not be so important, but together they add up. First, it has a six-speed gearbox with a closer spread of ratios and a taller top gear. Second, it weighs 70kg less than the more diminutive M coupe. And while the 911's 64-litre fuel tank is at best marginal in providing a reasonable touring range if you believe in tapping the car's performance, the BMW's puny

FACTFILE

	BMW Z3 M COUPE	PORSCHE 911 CARRERA
HOW MUCH?	£40,000-£42,000 (est)	£64,800
HOW FAST?		
0-30mph	2.1sec	1.8sec
0-60mph	5.1sec	4.6sec
0-100mph	12.0sec	10.5sec
30-70mph through gears	4.4sec	3.8sec
30-50mph (3rd)	3.3sec	3.3sec
50-70mph (4th)	4.6sec	3.9sec
Top speed	155mph*	174mph

*electronically limited

	BMW Z3 M COUPE	PORSCHE 911 CARRERA
HOW THIRSTY?		
Urban	17.0mpg	16.0mpg
Extra urban	35.7mpg	32.8mpg
Combined	25.4mpg	23.7mpg
HOW BIG?		
Length	4025mm	4430mm
Width	1740mm	1765mm
Height	1280mm	1305mm
Wheelbase	2459mm	2350mm
Weight	1390kg	1320kg
Fuel tank	51 litres	64 litres
ENGINE		
Layout	6 cyls in line, 3201cc	6 cyls horizontally opposed, 3387cc
Max power	321bhp at 7400rpm	296bhp at 6800rpm
Max torque	258lb ft at 3250rpm	258lb ft at 4600rpm
Specific output	100bhp per litre	87bhp per litre
Power to weight	231bhp per tonne	224bhp per tonne
Installation	Longitudinal, front, rear-wheel drive	Longitudinal, rear, rear-wheel drive
Made of	Aluminium head, iron block	Alloy heads and block
Bore/stroke	86.4/91mm	96/78mm
Compression ratio	11.3:1	11.3:1
Valve gear	4 per cyl, dohc, variable exhaust and inlet valve timing	4 per cyl, dohc, variable valve timing
Ignition and fuel	MS S50 digital management, electronic injection	DME engine management, electronic injection
GEARBOX		
Type	5-speed manual	6-speed manual
Ratios/mph per 1000rpm	1st 4.21/5.5	1st 3.82/5.6
	2nd 2.49/9.4	2nd 2.20/9.8
	3rd 1.66/14.0	3rd 1.52/14.2
	4th 1.24/18.8	4th 1.22/17.5
	5th 1.00/23.2	5th 1.02/21.0
		6th 0.84/25.7
Final drive	3.15	3.44
SUSPENSION		
Front	Struts, coil springs, anti-roll bar	Struts, coil springs, anti-roll bar
Rear	Semi-trailing arms, coil springs, anti-roll bar	Multi-link, coil springs, anti-roll bar
STEERING		
Type	Rack and pinion, power assisted	Rack and pinion, power assisted
Lock to lock	3.2 turns	3.0 turns
BRAKES		
Front	315mm vented discs	318mm vented discs
Rear	312mm vented discs	299mm vented discs
Anti-lock	Standard	Standard
WHEELS AND TYRES		
Size	7.5Jx17in (f), 9Jx17in (r)	7Jx17in (f), 9Jx17in (r)
Made of	Aluminium alloy	Cast alloy
Tyres	Michelin Pilot SX 225/45 ZR17 (f) 245/40 ZR17 (r)	Continental SportContact 205/50 ZR17 (f) 255/40 ZR17 (r)

Looks will make BMW a winner for many; Porsche is ultimately better car

51-litre capacity is woefully inadequate, despite slightly better fuel consumption. The 911's 0.30 drag co-efficient is significantly more slippery than the Z3's 0.37.

You've got to ask the obvious: do these differences, and the 911's bigger two-plus-two cabin, make the Porsche worth another £24,000 on top of the M coupe's anticipated £41,000? Depends on what you want from a car.

Porsche, this is a raw beast – loud, aggressive and extroverted. Not quite as quick, it's true, though in the real world the difference is minimal.

The point here is that the M coupe actually feels and sounds more explosive than the civilised 911. Does perception become reality? Not quite. Stir the Porsche and its supremely smooth and seamless engine delivers superior performance,

BMW cabin is cramped for tall drivers; dash comes straight from roadster

Porsche two-plus-two is roomier and lighter, but seats are too narrow

I've done enough miles in the new 911 to know it's finally scare-free, a wonderfully talented GT that's also capable of turning into a fully fledged sports car when the driver's mood and the road demands.

There are those, however, who find this new-found refinement excessive, as if some of the 911-ness has been engineered out of the new car. For some, its gigantic performance and vastly improved handling and roadholding is not enough.

The M coupe might have been created specifically for these doubters. Maybe, in truth, they were exactly the people the BMW engineers had in mind when they developed the M coupe. By the standards of the

via a more satisfying and perfectly weighted gearchange, and an intoxicating howl from the tail that's more than a match for the rumbling extravagancies of the aurally more highly strung BMW.

The Porsche's steering is lighter. The constant writhing that some people enjoy has been filtered out, leaving only the information the driver really needs. On the other hand, the M coupe's steering is heavier, less direct and communicative, despite the huge advance over the roadster in feel and agility. Yet, it's still the Porsche that is the more nimble and accurate.

This is the first new 911 I've driven on the standard 17in wheels, and the ride is even

Is 911 worth £24,000 more than Z3 as mid-life healing for the well heeled?

more supple than on the optional 18s. Brilliant. The M coupe, too, rides superbly for a car on such low-profile tyres, but it can't match the compliance or body control of its rival and is occasionally caught out over manhole covers or deep potholes.

It's easier to drive the 911 fast on twisty roads. Maybe that fact, too, will make the BMW more desirable for some people who desire raw exposure to the car's talents. Not that it is ever rough or crude.

At first, judging by its looks and its uneven idle, it seems entirely possible that the Z3 coupe might be both rough and crude. But more experience shows that apart from some low-speed driveline shunt, it is remarkably refined compared with the roadster version.

Both these coupes offer huge performance, responsive handling, fabulous braking ability and massive grip, at least in most circumstances.

The BMW can spin both rear wheels exiting a tight first or second-gear corner under full noise even on a dry road. Back off in the 911 and the tail still edges out, yet so benign is its behaviour that there will be those who prefer the M coupe's flagrant power oversteer.

For both, 125mph autobahn cruising is a doddle. At these speeds and above, the BMW has a small but clear stability advantage over the Porsche.

The M coupe's cabin is tiny, and forward of the seats is identical to the roadster's. So is the same awkward driving position – cramped for the tall – the non-adjustable steering and gaudy instruments.

Fortunately the BMW has the better seats, beautifully comfortable and dished buckets that hold you in place and are fully supportive without the driver being conscious of their high quality.

In contrast the bucket backs on the 911 are too narrow, especially at shoulder height. But the pedals and steering wheel are square on, so the Porsche locates the driver more comfortably than the BMW in an open, airy cabin that's for the most part tastefully trimmed in leather, the exception being some cheap plastics.

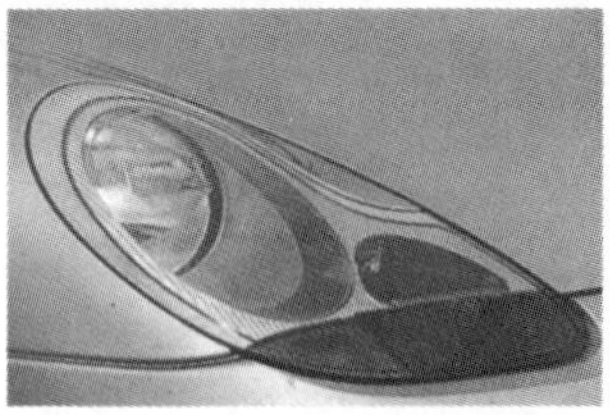

Mean squint of M coupe is in keeping with BMW's less sophisticated feel

These cars are a great way to forget the mid-life crisis. They provide enough drama for any well-heeled automotive enthusiast suffering the pangs of middle age.

That doesn't stop the 911 from being the better car. More refined it may be, but it is also more multi-talented. But that doesn't stop us from understanding the overt appeal of the brutal M coupe.

Buy one or the other and save on the psychiatrist's fees.

Coupe more refined than roadster

Porsche has better body control

BMW M ROADSTER

At last, a wolf in Z3 clothing

BY ANDREW BORNHOP
PHOTOS BY JEFF ALLEN

ARE YOU A bit dissatisfied with BMW's 4-cylinder Z3? Me too. It's eye-catchingly muscular, but the tinny-sounding 1.9-liter engine doesn't pack enough punch to back up the car's powerful looks...which is counter to my tastes and not exactly what you'd expect of BMW. Give me a sleeper car over a poser any day. A Miata, for instance, which is significantly less expensive but equally at home on a twisting back road.

BMW's Z.3 2.8 is another story. Its 189-bhp inline-6—complemented by beefier suspension bits and a richer-feeling interior—creates the refined Z3 that BMW needed to build. If there's one thing the Z3 2.8 lacks, however, it's this: a voracious appetite satisfied only by a diet rich in throttle usage and high in lateral g's. In short, the Z3 2.8 is more a polished GT, less a feisty sports car.

Now things have changed. With the April arrival of the M roadster, BMW will finally bring a Z3 to the American market that makes rabid sports-car enthusiasts stand back and say, "Now that's more like it." And if there's any doubt about the M roadster living up to its "mini Cobra" appearance, consider this—it's the quickest BMW ever sold in the U.S.

Glance at our data panel and see why—this South Carolina-built M roadster is basically a potent M3 wearing impressively stiff (and light) Z3 roadster clothing. Underhood is the M3's twincam 3.2-liter inline-6, which sends its 240 bhp to the rear wheels via the M3's 5-speed manual gearbox and a limited-slip differential adorned with enough low-flying cooling fins to turn a potato into a dozen French fries.

Starchy snacks aside, the 3,085-lb. M roadster also benefits from the M3's monstrous vented disc brakes (more than a foot in diameter) at each wheel, plus a sport-tuned suspension that lowers the ride height by an inch. The rear suspension isn't the M3's multilink arrangement; rather, it's a version of the Z3 2.8's trailing arms, beefed up to handle the significant stresses generated by BMW M GmbH's powerplant.

Compared with the Z3 2.8, the M roadster has firmer springing and damping in front, and a 1-mm-smaller anti-roll bar for improved turn-in response. In back, the springs have actually been softened for better off-the-line traction, and wheel hop is checked by firmer damping. To further reduce understeer, BMW has switched to a 3.5-mm-larger rear bar, which, as in the front, links to the suspension via balljoints for crisp response. What's more, the M roadster wears tire rubber galore—225/45ZR-17s in front and

■ BMW doles out the nostalgia, courtesy of two-tone coloring and liberal use of chrome. Note the passenger-airbag deactivation switch, just above the shifter that illuminates when headlights are on.

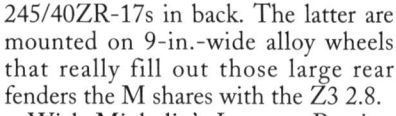

245/40ZR-17s in back. The latter are mounted on 9-in.-wide alloy wheels that really fill out those large rear fenders the M shares with the Z3 2.8.

With Michelin's Laurens Proving Grounds in South Carolina kindly put at our disposal, we probed this Estoril Blue M roadster for what it's worth. Namely, a sprint to 60 mph in a fleet 5.2 seconds and a quarter-mile blast in 14 flat at a heady 98.2 mph. Bye, bye, Boxster. Find a quicker BMW in our Road Test Summary. You can't.

Subjectively, the reassuringly firm brakes of the M roadster get top marks. No surprise, our objective numbers back up that grade. This car stops as quickly as a Corvette, which is to say it can haul itself down from 80 mph in just a speck over 200 ft. Most impressive. And 4-channel ABS removes any hint of drama from our simulated panic stops.

The slalom is trickier. Because the M roadster reacts so sharply to initial steering inputs (the variable-ratio rack comes straight from the M3), it's deceptively easy to crank in too much steering at the first pylon and start the car off on a path that's far curvier (and full of greater lateral forces) than need be. But with minimum steering input and a straight-as-possible path, the M roadster slices through the cones at an outstanding 65.3 mph.

It's the immediate responsiveness, the agility, that makes the M roadster a delight on Michelin's handling

THE COMPETITION

Chevrolet Corvette Convertible

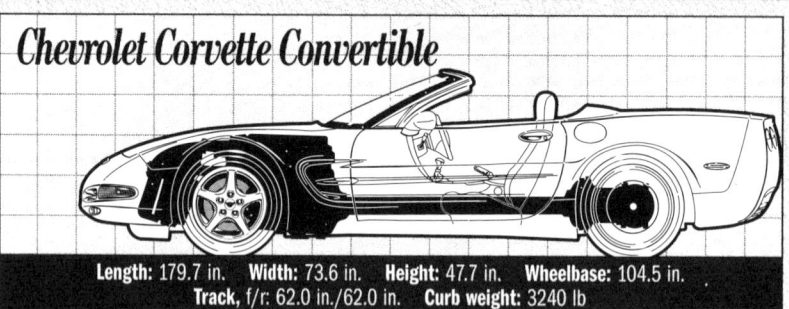

Length: 179.7 in. **Width:** 73.6 in. **Height:** 47.7 in. **Wheelbase:** 104.5 in.
Track, f/r: 62.0 in./62.0 in. **Curb weight:** 3240 lb

■ Will many people cross-shop the M roadster and the new Corvette convertible? Probably not. But as a pair of high-performance roadsters listing for less than 45 grand, a comparison makes sense. The much-larger Vette, despite having a luscious aluminum small-block V-8, is less brutal, less rambunctious. At the same time, it manages to get its nose ahead of the Bimmer's in the quarter mile. That's what 345 bhp will do for you, mounted in a sleek open-top chassis that's just as stiff as the new M roadster's. *(Tested: 9/97)*

Current list price	$44,425
Engine	ohv 5.7-liter V-8
Horsepower	345 bhp @ 5600 rpm
Torque	350 lb-ft @ 4400 rpm
Transmission	6-speed manual
0–60 mph	5.2 sec
Braking, 60–0 mph	118 ft
Lateral accel (200-ft skidpad)	na
EPA city/highway	18/28 mpg

Porsche Boxster

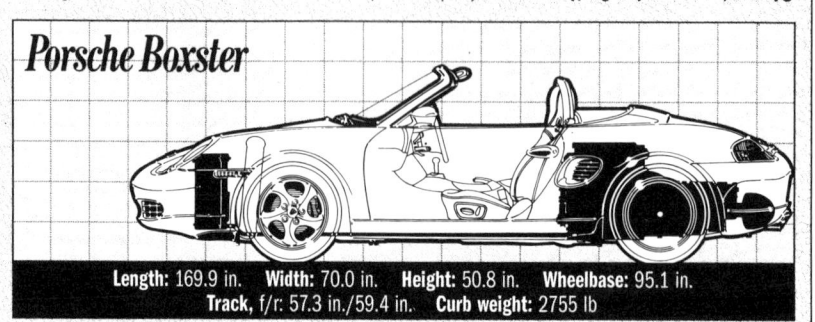

Length: 169.9 in. **Width:** 70.0 in. **Height:** 50.8 in. **Wheelbase:** 95.1 in.
Track, f/r: 57.3 in./59.4 in. **Curb weight:** 2755 lb

■ It's just about impossible to get a Boxster out of shape. With lots of dialed-in understeer and a rear end that's firmly planted to the ground, this Porsche remains impressively composed in a variety of twisty-road environments. The M roadster, on the other hand, demands more of its driver. And in acceleration, Munich's monster easily outdistances Stuttgart's stallion. But before too long, Porsche will come out with the Boxster S, rumored to have 300 bhp. The race goes on... *(Tested: 3/97)*

Current list price	$41,000
Engine	dohc 2.5-liter flat-6
Horsepower	201 bhp @ 6000 rpm
Torque	181 lb-ft @ 4500 rpm
Transmission	5-speed manual
0–60 mph	6.1 sec
Braking, 60–0 mph	120 ft
Lateral accel (200-ft skidpad)	0.93g
EPA city/highway	19/26 mpg

1998 BMW M ROADSTER

MANUFACTURER
BMW Manufacturing Corp.
P.O. Box 11000
Spartanburg, South Carolina 29304-4100

PRICE
List price..................... est $41,900
Price as tested .... est $42,853
    Price as tested includes std equip. (dual airbags, ABS, air cond, cruise control, heated pwr seats, AM/FM stereo/cassette; pwr windows, mirrors & top; leather upholstery, central locking, rollover protection bars, limited-slip differential, passenger airbag deactivation switch), luxury tax (est $383), dest charge ($570).

0–60 mph................ 5.2 sec
0–¼ mi .... 14.0 sec
Top speed .... 137 mph*
Skidpad 0.89g
Slalom .... 65.3 mph
Brake rating .... excellent

TEST CONDITIONS
Temperature................................ 53° F
Wind .... calm
Humidity..................................... na
Elevation................................. 500 ft

ENGINE
Type................... cast-iron block, aluminum head, **inline-6**
Valvetrain .... dohc 4 valve/cyl
Displacement ..... 192 cu in./3152 cc
Bore x stroke .... 3.40 x 3.53 in./
    86.4 x 89.6 mm
Compression ratio............. 10.5:1
Horsepower
  (SAE) ...... **240 bhp @ 6000 rpm**
Bhp/liter 76.1
Torque .... **236 lb-ft @ 3800 rpm**
Maximum engine speed ..... 6800 rpm
Fuel injection .... elect. sequential port
Fuel ..... prem unleaded, 91 pump oct

CHASSIS & BODY
Layout .... **front engine/rear drive**
Body/frame .... unit steel
Brakes
  Front .... 12.4-in. vented discs
  Rear .... 12.3-in. vented discs
Assist type .... vacuum; ABS
Total swept area .... 491 sq in.
Swept area/ton........... 294 sq in.
Wheels.................... cast alloy;
  17 x 7½ f, 17 x 9 r
Tires .... Michelin Pilot SX MXX3;
  225/45ZR-17 f, 245/40ZR-17 r
Steering ..... **rack & pinion**, pwr assist
  Overall ratio 17.8:1
    (variable, mean ratio)
  Turns, lock to lock 3.2
  Turning circle .... 34.1 ft
Suspension
  Front .... **MacPherson struts, L-shaped lower arms,**
    coil springs, tube shocks,
    anti-roll bar
  Rear .... **semi-trailing arms,**
    coil springs, tube shocks,
    anti-roll bar

DRIVETRAIN
Transmission.................................... **5-speed manual**

Gear	Ratio	Overall ratio	(Rpm) Mph
1st	4.21:1	13.59:1	(6500) 35
2nd	2.49:1	8.04:1	(6500) 60
3rd	1.66:1	5.36:1	(6500) 90
4th	1.24:1	4.01:1	(6500) 120
5th	1.00:1	3.23:1	(5980) 137*

Final drive ratio 3.23:1
Engine rpm @ 60 mph in 5th 2600
*Electronically limited.

GENERAL DATA
Curb weight.............. **est 3085 lb**
Test weight .... est 3335 lb
Weight dist (with
  driver), f/r, % .... est 51/49
Wheelbase................... 96.8 in.
Track, f/r .... 55.0 in./58.7 in.
Length..................... **158.5 in.**
Width...................... **68.5 in.**
Height..................... **49.8 in.**
Ground clearance................. na
Trunk space................ 6.2 cu ft

MAINTENANCE
Oil/filter change .... 7500 mi/7500 mi
Tuneup .... 30,000 mi
Basic warranty...... 48 mo/50,000 mi

ACCOMMODATIONS
Seating capacity 2
Head room .... 37.0 in.
Seat width............... 2 x 20.5 in.
Leg room.................... 42.5 in.
Seatback adjustment .... 22 deg
Seat travel.................... 8.0 in.

INTERIOR NOISE
Idle in neutral............... 54 dBA
Maximum in 1st gear.......... 85 dBA
Constant 50 mph............. 73 dBA
  70 mph .... 76 dBA

INSTRUMENTATION
160-mph speedometer, 8000-rpm tach, coolant temp, oil temp, fuel level, analog clock

ACCELERATION
Time to speed	Seconds
0–30 mph	1.8
0–40 mph	2.9
0–50 mph	4.0
0–60 mph	5.2
0–70 mph	7.2
0–80 mph	9.0
0–90 mph	11.6
0–100 mph	14.6

Time to distance
  0–100 ft. 2.8
  0–500 ft. 7.5
  0–1320 ft (¼ mi): 14.0 @ 98.2 mph

FUEL ECONOMY
Normal driving.......... est 22.0 mpg
EPA city/highway........ est 20/27 mpg
Cruise range .... est 275 miles
Fuel capacity .... 13.5 gal.

BRAKING
Minimum stopping distance
  From 60 mph.................. 116 ft
  From 80 mph.................. 203 ft
Control .... excellent
Pedal effort for 0.5g stop........... na
Fade, effort after six 0.5g stops from
  60 mph .... na
Brake feel .... excellent
Overall brake rating .... excellent

HANDLING
Lateral accel (200-ft skidpad) ... 0.89g
Balance............. mild understeer
Speed thru 700-ft slalom .... 65.3 mph
Balance .... neutral
Lateral seat support.......... excellent

Test Notes...

■ The M roadster launches with authority and beats an M3 sedan in the quarter mile. But the sedan reaches 100 mph first, likely because of its better aerodynamics.

■ Want to know what a firm brake pedal should feel like? Drive this M roadster. BMW and Porsche ought to teach other carmakers about proper brake feel.

■ With its short wheelbase and high power, the M roadster can be teased into gratifying power oversteer on the skidpad. But in steady-state cornering, it mildly understeers.

Subjective ratings consist of excellent, very good, good, average, poor; na means information is not available.

SCALE: 10 IN. (254mm) DIVISIONS
DRAWING BY TIM BARKER

Looking for a sure-fire way to spot an M roadster? It's the Z3 with a quad-tip exhaust, M badges, trunk-mounted license plate and unique wheels and tires that properly fill out the car's muscular flanks. Though M roadsters built for Europe have 321 bhp, the U.S. model pumps out a solid 240.

course. Any time you mix a short wheelbase (10 in. less than the M3's) with gobs of rear-drive power, you're bound to get a car that likes to slide its tail. The M roadster is no exception. Despite a high level of grip from the Michelin Pilots, the rear tires can (and will) break away when you're powering out of a corner or lifting off the throttle and turning in. Nothing spooky here, just enough drama to keep the driver attentive. And sitting so close to the rear wheels—as you do in this traditionally laid-out roadster—only enhances this pendulum feel.

Back to the phenomenal acceleration for a moment. With 240 bhp on tap, a broad band of torque (whose peak of 236 lb.-ft. is reached at 3,800 rpm) and five tightly spaced gear ratios, the M roadster really squirts. Need to pass somebody on the highway? The throaty-sounding inline-6 makes quick duty of it, pulling like a freight train to its 6,500-rpm redline. And around town—where most cars loaf along in 3rd gear—the M roadster cruises comfortably in 4th. This relates directly to the 3.23:1 final drive and gearing so low that 5th isn't even an overdrive. Clutch take-up, by the way, is second-nature, and the short-throw shifter moves firmly from gear to gear with only a bit more resistance than is felt in the Z3 2.8.

As is true of any BMW fit to wear Munich's motorsports badge, the M roadster is endowed with numerous unique pieces, some purely cosmetic, others not. These include new bumper caps, tricolor BMW M badges (in a variety of places), splashes of chrome trim (inside and out) and exclusive 17-in. alloy wheels. Moreover, the trunklid is different, carrying the BMW roundel on its upper surface and the license plate on its short vertical plane.

The M roadster's air dam has a larger opening because the German-built M engine (which is shipped in assembled form to BMW's plant in Spartanburg, South Carolina) requires more air than the 2.8. And its side ports serve as brake cooling ducts, not places to put foglights. The rear bumper, on the other hand, had to be altered to make room for the M roadster's dual-muffler/quad-tip exhaust, a feature that will appear on all future M cars. Because the passenger-side muffler occupies the space formerly reserved for the spare tire, BMW equips the M roadster with a tiny electric air pump and a fast-acting tire sealer that stow away neatly in the trunk. Also in the trunk is the battery, hung low between the mufflers in a box that is part of the car's unique trunk-floor stamping. This helps with weight distribution, which is pegged by BMW at a smart 51/49.

Other unique bits include a larger, more ovoid rearview mirror (which can get in the way when you're looking toward the apex of a right-hand corner), a pair of stylish but less effective (i.e., smaller) sideview mirrors, two-tone sport seats that offer superb lateral support (particularly for your upper back) and a new center console thats graced with six chrome-trimmed gauges and controls that give the car a warm, decidedly retro feel. That feeling is augmented by chrome trim around the shift boot and the car's excellent analog gauges, but there's nothing retro about the M roadster's standard level of equipment—it's loaded. A power top, heated power seats, one-touch side windows and air conditioning are all standard, as is beautifully stitched leather that covers much of the interior, even the new thick-rimmed 3-spoke steering wheel.

Taller drivers (those over 6 ft. 2 in. or so) will wish they could move the seat back an additional inch, but the M roadster is nevertheless a comfortable car for most, and its highway ride is on the firm side of comfortable—just what I like. One thing I don't like, however, are the car's limp seatbelt retractors. Twice I shut the door on the belts because they had not retracted properly.

People in the Snowbelt may think again about owning this powerful, wide-tired M roadster, primarily because it doesn't have traction control. It's in the works, says BMW, but it hasn't been tailored to the car yet. BMW says the challenge is in making a system that responds fast enough to catch the explosive wheelspin of this short-wheelbase roadster.

But this sort of concern doesn't detract from a car such as the Shelby Cobra. And it doesn't with the M roadster. This is a well-built weekend car that's refined enough to drive every day. Only 3,000 are planned for the U.S. in 1998, at an estimated price of $41,900 that includes free maintenance for three years or 36,000 miles. Your smiles are guaranteed to last a lot longer.